SCHOLASTIC

READ & SOLVE MATH MINI-BOOKS

by Tracy Jarboe and Stefani Sadler

NEW YORK • TORONTO • LONDON • AUCKLAND • SYDNEY
MEXICO CITY • NEW DELHI • HONG KONG • BUENOS AIRES

Teaching Resources

This book is dedicated to our newest generation of teachers, Marisa and Kelly.
We wish you many years of joyful teaching!

Cover design by Jason Robinson

Interior design by Sydney Wright

Cover and interior art by Stefani Sadler

ISBN–13: 978-0-439-52979-2
ISBN–10: 0-439-52979-4

Published by Scholastic Inc.

Printed in the U.S.A
2 3 4 5 6 7 8 9 10 31 14 13 12 11 10 09 08 07

Contents

Introduction

As classroom teachers, we have witnessed the variety of ways in which students respond to instructional techniques across the curriculum. We have found that one of our most effective methods of math instruction has been the use of interactive manipulative-based activities. Our favorite approach to this type of learning is student- or class-made math mini-books. For this reason we have created *Read & Solve Math Mini-Books*—a collection of 12 mini-books for young learners that introduce and reinforce essential mathematical skills.

Teachers are constantly learning about how best to reach students, and modifying and supplementing their curriculum accordingly. In an effort to help simplify this process, we have created a comprehensive resource that will enhance and extend the core concepts already being taught in your classroom—in a fun and engaging way! These mini-books also meet the standards set forth by the National Council of Teachers of Mathematics (NCTM). (Turn to page 5 to see how each mini-book connects to the current standards.) *Read & Solve Math Mini-Books* fosters the development of the following essential math skills:

Number Sense	Time
Addition	Days of the Week
Subtraction	Money
Shapes	Classification
Patterns	Graphing

There are many benefits to using mini-books in math instruction. In addition to their math focus, these mini-books also connect to your literacy instruction. The format of each mini-book provides many purposeful literacy experiences. As students read each mini-book, they have the opportunity to decode text, develop their word-recognition abilities, expand their vocabularies, and improve their fluency through repeated readings. Students also build their writing skills as they respond to the interactive text by writing and recording mathematical data.

These mini-books are easily incorporated into any classroom or curriculum. They can be utilized for whole- class or small-group instruction or for independent practice. Plus, the activities for each lesson are flexible so they can be adjusted to suit the needs of the children in your classroom. Teaching notes for each of the mini-books can be found on pages 7–18 and include objectives, tips, activities, extensions, and more. When the books are complete, you might place them in a learning center with complementary activities or literature to be revisited again and again. Or, send the mini-books home with students to read with family members!

Connections with the NCTM Standards

	Number & Operations	Algebra	Geometry	Measurement	Data Analysis & Probability	Problem Solving	Reasoning & Proof	Communication	Connections	Representation
I See Numbers	✓					✓		✓	✓	✓
Missing Numbers!	✓				✓	✓	✓	✓	✓	✓
One More Apple	✓				✓	✓	✓	✓	✓	✓
How Many in All?	✓				✓	✓	✓	✓	✓	✓
Five Pennies to Spend	✓				✓	✓	✓	✓	✓	✓
Shapes All Around	✓		✓			✓		✓	✓	✓
Super Shapes!	✓		✓			✓		✓	✓	✓
Patterns Everywhere	✓	✓	✓		✓	✓	✓	✓	✓	✓
Graph It!	✓				✓	✓		✓	✓	✓
Long and Short	✓			✓	✓	✓		✓	✓	✓
Around the Clock	✓			✓		✓		✓	✓	✓
We Eat Through the Week	✓			✓		✓		✓	✓	✓

How to Make the Mini-Books

Each of the mini-books is a snap to assemble. Your students' level of involvement will depend on their age and skill level. For younger students, you might want to copy, cut, and staple a book for each student prior to the lesson. For more advanced learners, simply copy the mini-book pages beforehand and then guide students as they assemble their own books. Follow the steps below to make each mini-book:

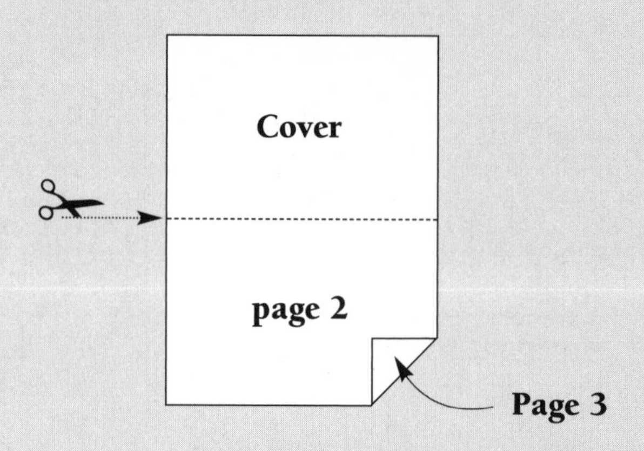

1. Copy the pages for the books onto standard 8½- by 11-inch paper, making the copies single-sided. (Each full-size book page contains two mini-book pages.)

2. Using the double-sided function of the copier, create double-sided copies. To do this, place the pages in the copier in the order in which they exist in this book. For example, the page containing the cover of the mini-book and page 2 should be followed by pages 1 and 3 so that pages 1 and 3 will copy directly behind the cover and page 2. (You might want to make a test copy to be sure the pages are aligned correctly.) Then, continue by adding pages 4 and 6 behind pages 1 and 3, followed by pages 5 and 7, and so on.

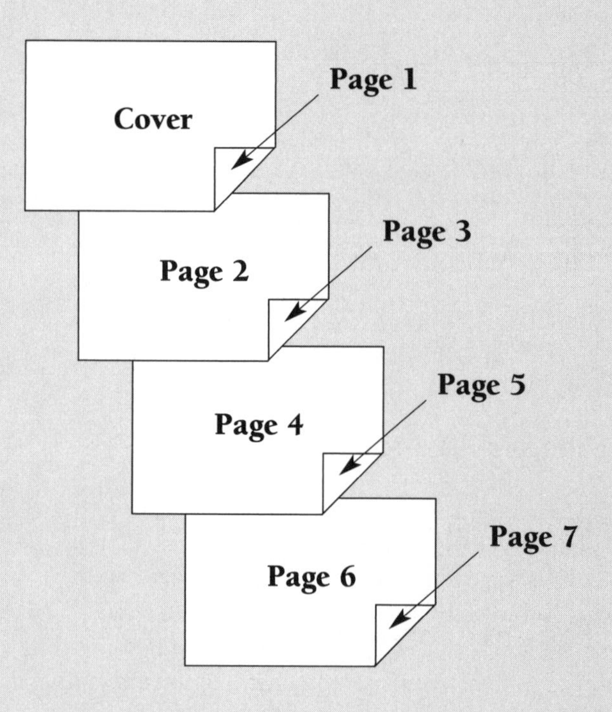

3. Cut the mini-book pages apart along the dashed lines.

4. Place the pages in numerical order, and then staple them along the mini-book's spine.

How to Use the Mini-Books

Number Sense

I See Numbers

Pages 20–25

In this mini-book, students practice identifying, counting, matching, and recording the numerals and number words for the numbers 1 through 10. Students need pencils, crayons, markers, or colored pencils to complete this activity.

Before Teaching

1. Copy the mini-book on pages 20–25 for each student.
2. Assemble a mini-book to preview with students by following the directions on page 6.
3. Label ten sentence strips with the numerals 1 through 10. Label another set of strips with the number words *one* through *ten* (in lowercase).

Getting Started

Randomly arrange the prepared sentence strips into two mixed-up columns. Label one *Numerals* and the other *Number Words*. Then invite students to correct the columns to match each numeral with its corresponding number word. For continued practice, ask volunteers to collect or identify objects in the classroom and match them to each number—for example: *two clocks, four windows, seven pencils,* and so on.

Making and Using the Mini-Book

1. Share your pre-assembled mini-book with students. Read the short number poem on each page and demonstrate how to trace the featured numeral and number word, and then add that same number of objects to the illustration.
2. Distribute pages 20–25 and help students assemble according to the directions on page 6.
3. Support students as they work through their mini-books. They may need help adding objects to the illustrations on each page. Then invite students to decorate their mini-books.
4. Provide time for children to share their books, as a whole-class read-aloud, with partners, or with family members at home.

Taking It Further

Provide students with additional practice working with numerals and number words by creating number-line and number-word scrambles. For numeral practice, create a number line with the numerals ordered incorrectly. Then, invite students to correct the mistakes and order the numerals appropriately. For practice with number words, write the words, or build them using letter tiles, and rearrange the letters within each word. Then, ask students to unscramble the letters and decode the word.

Number Sense

Missing Numbers!

Pages 26–29

In this mini-book, students will identify and order numbers 1 through 10. Students need pencils, crayons, markers, or colored pencils to complete this activity.

Before Teaching

1. Copy the mini-book on pages 26–29 for each student.
2. Assemble a mini-book to preview with students by following the directions on page 6.
3. Prepare a counting chart or number line with several numbers missing. Make a copy for each student.

Getting Started

With students, review the numbers 1 through 10. You might count the numbers forward or backward, or pose questions about which number comes before or after another number. Next, provide students with the prepared counting chart or number line that has several missing numbers. Invite students to complete the chart or number line by filling in the missing numbers. Have students work individually, with partners, or as a whole class, and then review the results.

Making and Using the Mini-Book

1. Share your pre-assembled mini-book with students. Read each page aloud and point out the places where numbers need to be added to the illustration.
2. Distribute pages 26–29 and help students assemble according to the directions on page 6.
3. Support students as they work through their mini-books. Then, invite them to further illustrate and/or decorate their mini-books.
4. Provide time for children to share their books, as a whole-class read-aloud, with partners, or with family members at home.

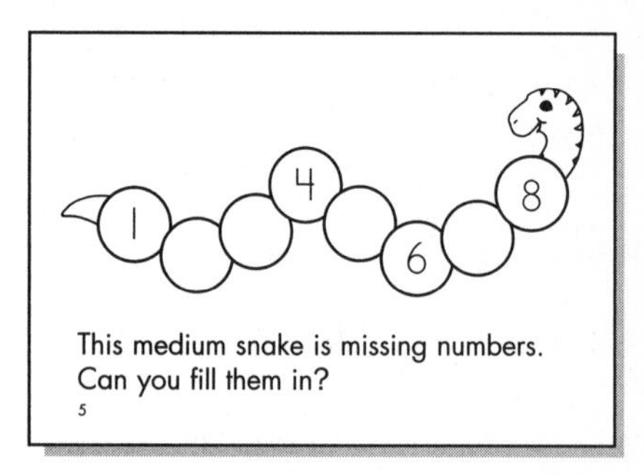

Taking It Further

Using index cards, create number cards for the numbers 1 through 10. Arrange the cards in sequential order, and as a class, count the numbers aloud. Then ask students to cover their eyes while you remove one or more of the number cards. Challenge students to uncover their eyes and identify the missing number(s) as quickly as possible!

Addition

One More Apple

Pages 30–35

In this mini-book, students read and solve "plus-one" addition problems using numbers 0 through 9, working with both numerals and number words. Students need pencils, crayons, markers, or colored pencils to complete this activity.

Before Teaching

1. Copy the mini-book on pages 30–35 for each student.
2. Assemble a mini-book to preview with students by following the directions on page 6.

Getting Started

Using chart paper or a whiteboard, create "plus-one" addition sentences such as the one shown here, to serve as models for completing the mini-book. Invite a volunteer to record the equation, *3 + 1 = 4*. Repeat until students have a solid grasp of the concept.

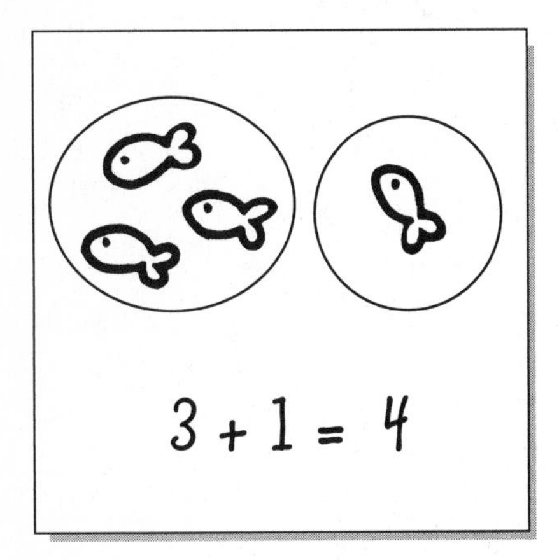

For further practice, encourage students to use class manipulatives and create their own addition sentences. Then go around the room while the class takes turns identifying one another's sentences.

Making and Using the Mini-Book

1. Share your pre-assembled mini-book with students and read the rhyme on each page. Then, model how to add the missing number word, trace the apple that is being added to the equation, and record the addition sentence that represents the illustration.
2. Distribute pages 30–35 and help students assemble according to the directions on page 6.
3. Support students as they work through their mini-books. Then, invite them to decorate their mini-books.
4. Provide time for children to share their books, as a whole-class read-aloud, with partners, or with family members at home.

Taking It Further

Play the game "Around the Apple Orchard." Students sit in a circle while you go around the "orchard" showing each student an addition flash card and allowing time for that child to provide the answer. For a more spirited version, assign one student to be the "farmer" and to stand behind another student seated in the circle. Show both students the addition flash card. The student who correctly answers the question faster continues on and stands behind the next student in the circle. The object is to make it all the way around the orchard without being beaten.

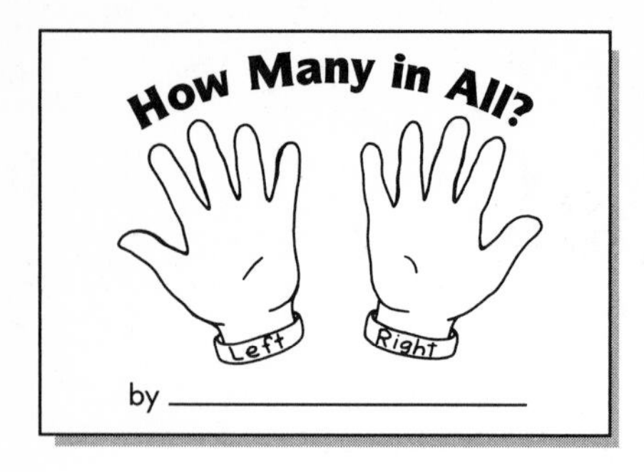

Addition

How Many in All?

Pages 36–41

In this mini-book, students solve addition problems using numbers 0 through 6. Students need scissors, glue sticks, pencils, crayons, markers, or colored pencils to complete this activity.

Before Teaching

1. Copy the mini-book on pages 36–41 for each student.
2. Assemble a mini-book to preview with students by following the directions on page 6.
3. Draw a large pair of hands on a sheet of chart paper and laminate.

Getting Started

Using the prepared chart, create addition problems such as the one shown here, to serve as models for completing the mini-book. You might wish to follow the format of the mini-book by creating problems that include buttons. Invite a volunteer to record the equation, $3 + 2 = 5$. Repeat until students have a solid grasp of the concept.

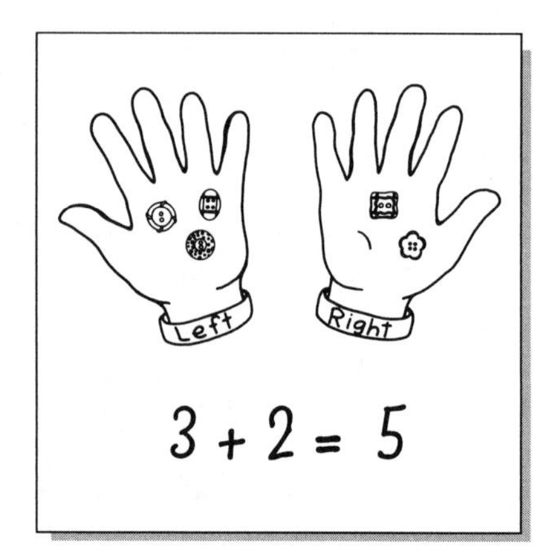

Making and Using the Mini-Book

1. Share your pre-assembled mini-book with students. Tell them that they should begin by cutting out the button manipulatives on page 11 of the mini-book. Explain that students will glue buttons into the hands according to the addition sentence on that page. Then, they will add the two groups of buttons together and record the answer on the blank line. Finally, on page 10, they will have the opportunity to create their own addition equation.
2. Distribute pages 36–41 and help students assemble according to the directions on page 6.
3. Support students as they work through their mini-books. Then, invite them to decorate their mini-books.
4. Provide time for children to share their books, as a whole-class read-aloud, with partners, or with family members at home.

Taking It Further

Practice a variety of math skills using a collection of old buttons:

- Create button patterns—for example, *brown button/black button*; or *circle button/square button/square button/circle button*.
- Sort buttons by color, shape, or another attribute—for example, *buttons with three holes* versus *buttons with four holes*.
- Count groups of buttons and order them from smallest to greatest—for example, *the red bowl has 10 buttons, the green bowl has 15 buttons, the blue bowl has 20 buttons. The blue bowl has the greatest number of buttons and the red bowl has the smallest number of buttons*.

Subtraction

Five Pennies to Spend

Pages 42–45

In this mini-book, students solve subtraction problems using numbers 0 through 5, working with both numerals and number words. Students need scissors, glue sticks, pencils, crayons, markers, or colored pencils to complete this activity.

Before Teaching

1. Copy the mini-book on pages 42–45 for each student.
2. Assemble a mini-book to preview with students by following the directions on page 6.

Getting Started

Using chart paper or a whiteboard, create subtraction problems such as the one shown here, to serve as models for completing the mini-book. You might wish to follow the format of the mini-book by creating problems that include pennies—for example, *I have three pennies in my bank, and I buy a stick of gum that costs one penny. How many pennies do I have left?* Invite a volunteer to record the equation, $3 - 1 = 2$. Repeat until students have a solid grasp of the concept.

Making and Using the Mini-Book

1. Share your pre-assembled mini-book with students. Tell them that they should begin by cutting out the penny manipulatives on page 7 of the mini-book. Then, read the rhyming subtraction problem on each page and ask students to fill in the answer by adding a number word to the end of the rhyme. Demonstrate how to illustrate the rhyme by gluing down the total number of pennies then crossing one out to show that it has been spent. Finally, model how to write the equation on the blank lines underneath the illustration.
2. Distribute pages 42–45 and help students assemble according to the directions on page 6.
3. Support students as they work through their mini-books. They may need help cutting, counting, and gluing pennies. Then, invite students to decorate their mini-books.
4. Provide time for children to share their books, as a whole-class read-aloud, with partners, or with family members at home.

Taking It Further

Set up a center as a store with common items, or things based on a theme, that all have price tags attached. Price the items to match student ability—a pencil might be 5 cents, a notepad might be 8 cents, and so on. Students begin with a given amount—for example, 10 cents—then decide what they would like to "buy" and subtract its value. Students might solve the problems using real or manipulative pennies or by writing them on a sheet of paper. Encourage students to work with partners and correct one another's work. Change the items and prices often for more practice. As students become more proficient, increase the difficulty of the prices.

Geometry

Shapes All Around

Pages 46–49

In this mini-book, students identify, create, and count the following shapes: circles, squares, rectangles, and triangles. Students will also encounter numbers, color words, and basic sight words. Students need pencils, crayons, markers, or colored pencils to complete this activity.

Before Teaching

1. Copy the mini-book on pages 46–49 for each student.
2. Assemble a mini-book to preview with students by following the directions on page 6.

Getting Started

Start a class discussion about shapes. Draw or show students a square, rectangle, triangle, and circle. Guide students to identify and describe each shape, recording ideas on chart paper or a whiteboard, highlighting the following attributes:

Square: 4 sides (all equal length), 4 corners
Rectangle: 4 sides (2 long, 2 short), 4 corners
Triangle: 3 sides, 3 corners
Circle: no straight lines, no corners

Also, draw attention to each shape word, as students will be writing these words in their mini-books.

Making and Using the Mini-Book

1. Share your pre-assembled mini-book with students and tell them that the sentences on each page are tasks that they need to complete. Read each sentence aloud and demonstrate how to trace the shape word, the shapes, and the numbers, and count and record the total number of shapes. You might want to review key vocabulary such as *trace, word, numbers, how many, all,* and the shape and color words. Also, point out that students will have a chance to draw two of each shape on the last page of their mini-books.
2. Distribute pages 46–49 and help students assemble according to the directions on page 6.
3. Support students as they work through their mini-books. Then, invite them to decorate their mini-books.
4. Provide time for children to share their books, as a whole-class read-aloud, with partners, or with family members at home.

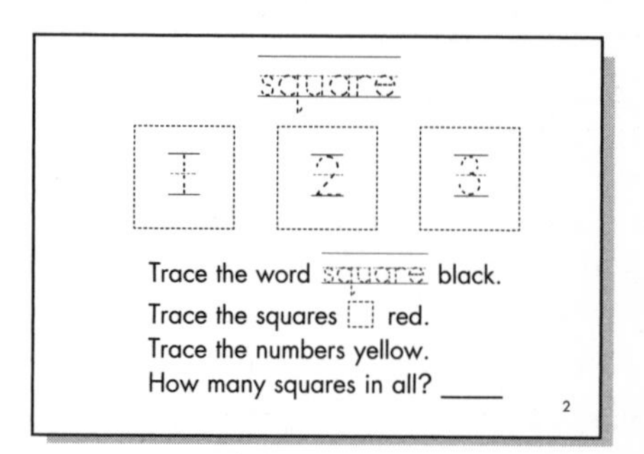

Taking It Further

- Using the four shapes introduced in this mini-book, ask students to draw real-life objects from each shape. For example, a circle can become a sun, a square or rectangle can become a television, a triangle can become an ice cream cone. Encourage students to be creative!
- Gather a collection of shapes such as erasers, balls, rulers, game pieces, manipulatives, or illustrations. Have children sort and classify the shapes or objects by attribute. You might ask students to explain and/or record their sorting rules to share with the class.

Geometry

Super Shapes!

Pages 50–53

In this mini-book, children will identify, create, and count the following shapes: ovals, crescents, stars, and hearts. Students will also encounter numbers, color words, and basic sight words. Students need pencils, crayons, markers, or colored pencils to complete this activity.

Before Teaching

1. Copy the mini-book on pages 50–53 for each student.
2. Assemble a mini-book to preview with students by following the directions on page 6.

Getting Started

Start a class discussion about shapes. Draw or show students an oval, crescent, star, and heart. Guide students to identify and describe each shape, recording ideas on chart paper or a whiteboard, highlighting the following attributes:

Oval: no straight lines, no corners
Crescent: 2 points, 2 curved sides
Star: 5 points, 10 sides
Heart: 2 points, 2 curved sides

Also, draw attention to each shape word, as students will be writing the words in their mini-books.

Making and Using the Mini-Book

1. Share your pre-assembled mini-book with students and tell them that the sentences on each page are tasks that they need to complete. Read each sentence aloud and demonstrate how to trace the shape word, the shapes, and the numbers, and count and record the total number of shapes. You might want to review key vocabulary such as *trace, word, numbers, how many, all,* and the shape and color words. Also, point out that students will have a chance to draw two of each shape on the last page of their mini-books.
2. Distribute pages 50–53 and help students assemble according to the general directions on page 6.
3. Support students as they work through their mini-books. Then, invite them to decorate their mini-books.
4. Provide time for children to share their books, as a whole-class read-aloud, with partners, or with family members at home.

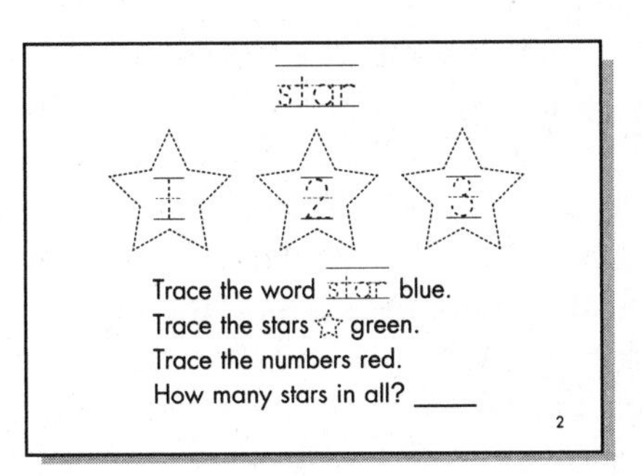

Taking It Further

Invite students to draw the shapes (or provide precut patterns and have students trace) on construction paper. Students can then cut out the shapes and decorate each using crayons, markers, tissue paper, glitter, and other craft supplies. These shapes can then be strung together to make a shape mobile!

Pre-Algebra

Patterns Everywhere

Pages 54–59

In this mini-book, students identify and extend patterns. Students need pencils, crayons, markers, or colored pencils to complete this activity.

Before Teaching

1. Copy the mini-book on pages 54–59 for each student.
2. Assemble a mini-book to preview with students by following the directions on page 6.

Getting Started

Start a class discussion about patterns. Invite a few volunteers up to the front of the room and arrange them to represent AB, ABB, and ABC patterns—for example, boy/girl, glasses/no glasses, tie shoes/non-tie shoes, pants/shorts/shorts, blonde/brunette/black. Ask students to identify and extend the patterns created by their classmates. Reinforce the terms *AB pattern*, *ABB pattern*, and *ABC pattern*, and explain that to extend a pattern means to continue it. Finally, create a whole-class pattern with students—for example, hands up/hands down, clap/snap/snap, stand/sit/kneel.

Making and Using the Mini-Book

1. Share your pre-assembled mini-book with students and tell them that each page contains an illustrated pattern that they need to identify and extend. Review each page, explaining to students that they will work with AB, ABB, and ABC patterns. On the last three pages of the mini-book, students will have a chance to create their own AB, ABB, and ABC patterns.
2. Distribute pages 54–59 and help students assemble according to the directions on page 6.
3. Support students as they work through their mini-books. They may need help identifying and extending the patterns on each page. Then invite students to decorate their mini-books.
4. Provide time for children to share their books, as a whole-class read-aloud, with partners, or with family members at home.

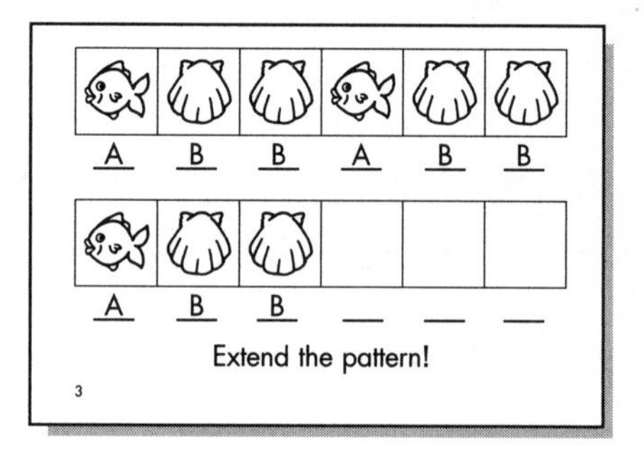

Taking It Further

Use any type of manipulative or other classroom material to encourage students to create and extend their own patterns. Increase the level of difficulty as children progress—AB patterns, to ABB patterns, to ABC patterns. For example, red cube/blue cube, pencil/crayon/crayon, circle/square/triangle.

Graphing

Graph It!

Pages 60–65

In this mini-book, students represent data using bar graphs. Students need pencils, crayons, markers, or colored pencils to complete this activity.

Before Teaching

1. Copy the mini-book on pages 60–65 for each student.
2. Assemble a mini-book to preview with students by following the directions on page 6.
3. Using chart paper or a whiteboard, prepare a simple bar graph with at least three columns and laminate. Students can then use a dry-erase marker, sticky notes, or name tags affixed with sticky tack or tape to record information.

Getting Started

Explain to students that many different types of graphs can be made to represent all sorts of data. Then introduce the steps of creating a class bar graph to answer the question, *"Which ice cream flavor do you like best?"* (1) Create a title for the graph. (2) Label three columns at the base of the graph *Vanilla, Chocolate,* and *Strawberry* (or three other flavors). (3) Label the vertical axis with an appropriate numeric range—for example, with a class of 20 students you might create a range of 0–15 as you most likely won't have one flavor that receives more than 15 votes. (4) Finally, let students fill in a bar on the graph above their favorite flavor using markers, sticky notes, or name tags.

Making and Using the Mini-Book

1. Share your pre-assembled mini-book with students. Explain that each two-page spread contains an illustration of animals in a particular setting on the left-hand page, and a graph on the right-hand page. Model the process of counting the number of each type of animal, recording that data in the accompanying graph, and answering three questions about the data. You might also demonstrate how to fill in the bar graph appropriately, encouraging students to color each bar neatly and carefully.
2. The last page of the mini-book contains a blank graph that students create individually. Ask students to survey a small group of students as the range is labeled 0–5. Some topics they might graph are eye color, pets, family members, and so on.
3. Distribute pages 60–65 and help students assemble according to the directions on page 6.
4. Support students as they work through their mini-books. They may need help counting information and filling in each graph. Then, invite students to decorate their mini-books.
5. Provide time for children to share their books, as a whole-class read-aloud, with partners, or with family members at home.

Taking It Further

Give each student a small amount of different-colored or -shaped cereal, pasta, or other classroom manipulative. Ask students to sort items by attributes such as color, shape, size, and so on. As a class, discuss an appropriate title, numeric range, and column labels. Then have students create a bar graph using the information. (Copy the blank graph from page 12 of the mini-book for each student to use.) After making their graphs, ask students to create two or three questions about their data and trade with a classmate to answer each other's questions!

Measurement

Long and Short

Pages 66–69

In this mini-book, students compare and create objects based on their length. Students need glue, pencils, crayons, markers, or colored pencils to complete this activity.

Before Teaching

1. Copy the mini-book on pages 66–69 for each student.
2. Assemble a mini-book to preview with students by following the directions on page 6.
3. Cut a group of long strands of string and a group of short strands of string. Create enough so that each student can have seven long strands and seven short strands. (Or have students draw the strings into the illustrations.)

Getting Started

Collect common classroom objects such as erasers, pencils, crayons, scissors, and so on. Pair shorter objects with longer objects and ask students to identify which is longer and which is shorter. (You may wish to extend this activity by actually measuring the objects with a ruler.) Next, invite two students to the front of the room. Ask the rest of the class questions—for example, *Who has longer hair? Who has shorter hair? Which child is wearing a long-sleeve shirt? Which child is wearing a short-sleeve shirt?* Continue making short and long comparisons until the students have had sufficient practice.

Making and Using the Mini-Book

1. Share your pre-assembled mini-book with students and tell them that each sentence is a task they need to complete. Read each sentence aloud and review new and/or challenging vocabulary. Then demonstrate how to glue long and short strands of string (or draw them) onto the illustration.
2. Distribute pages 66–69 and help students assemble according to the directions on page 6.
3. Support students as they work through their mini-books. They may need help reading sentences and gluing the long and short strands of string. After the glue dries, invite students to further illustrate and/or decorate their mini-books.
4. Provide time for children to share their books, as a whole-class read-aloud, with partners, or with family members at home.

Taking It Further

Read *Tails* by Matthew Van Fleet (Silver Whistle, 2003) to your class. Then, give each student a piece of adding-machine tape or a sentence strip onto which they will draw an animal tail. Encourage students to draw a specific animal's tail, either long or short, and to color it as accurately as possible. Then have students cut out the tails. Invite them to sort their tails into two groups: long or short. Within the two groups, help them order the tails by length from longest to shortest tail. Then, combine the two groups for a whole-class collection. You might even hang the tails on a bulletin board and invite students to write or dictate stories about their animals.

Time

Around the Clock

Pages 70–76

In this mini-book, students explore time to the nearest hour by recording the time of daily events on digital and analog clocks. Students need pencils, crayons, markers, or colored pencils to complete this activity.

Before Teaching

1. Copy the mini-book on pages 70–75 for each student.
2. Assemble a mini-book to preview with students by following the directions on page 6.
3. Copy sets of the analog and digital clocks from page 76. Create hour pairs from 8:00 a.m. to 8:00 p.m.

8:00

4. Copy several additional blank analog clocks to use during the Getting Started portion of the lesson.
5. Label sentence strips with daily events—for example, School Starts, Reading, Math, Writing, Lunch, School Ends, Dinner, Bedtime, and so on.

Getting Started

Begin a class discussion about daily schedules. What time do students wake up in the morning and go to school? What time are class activities such as reading, recess, math, lunch, and so on? What time do students leave at the end of the school day, do homework, eat dinner, and go to bed at night? Guide students toward using the nearest hour when discussing these events—for example, *We have math at 10:00 in the morning* or *We have reading at 1:00 in the afternoon.*

Model and explain the process of recording time to the nearest hour on analog clocks by drawing the hour and minute hands. Next, record various times on the photocopied blank analog clocks and ask students to identify them.

Finally, as a class create a sample daily schedule using the labeled sentence strips and digital and analog clock hour pairs. Invite students to find the matching clocks and place them next to the appropriate class activity.

9:00	School Starts
10:00	Math
11:00	Recess
12:00	Lunch
1:00	Reading
2:00	Writing
3:00	School Ends

Making and Using the Mini-Book

1. Begin by sharing your pre-assembled mini-book with students. As you read aloud the sentence on each page, point out the time written in word form—for example, *four o'clock*. Explain to students that they will then re-create that time on digital and analog clocks on the same page.
2. Distribute pages 70–75 and help students assemble according to the general directions on page 6.
3. Support students as they work through their mini-books, particularly with drawing the hour and minute hands. Remind them to double-check each page so that all time representations match exactly. Then, invite students to decorate their mini-books.
4. Provide time for children to share their books, as a whole-class read-aloud, with partners, or with family members at home.

Taking It Further

Create a simple song or chant to reinforce the concept of time throughout the day. For example, use the tune of "London Bridge" to help students with daily tasks:
It's time to put our work away, work away, work away.
It's time to put our work away, it's 3:00!

Time

We Eat Through the Week!

Pages 77–80

In this mini-book, students identify and order the days of the week. Students need pencils, crayons, markers, or colored pencils to complete this activity.

Before Teaching

1. Copy the mini-book on pages 77–80 for each student.
2. Assemble a mini-book to preview with students by following the directions on page 6.
3. Label flashcards or sentence strips with each day of the week.

Getting Started

Launch a class discussion about the days of the week. Ask students a few questions—for example, *What is the first day of the week? What is the seventh day of the week?* Using the prepared day-of-the-week sentence strips, invite students to place them in the correct sequence from Sunday through Saturday. Discuss ordinal numbers and demonstrate how to record them; *1st*, *2nd*, *3rd*, and so on, as students will be writing these numbers in their mini-books. Next, ask students to cover their eyes while you mix up the order of the cards. When they uncover their eyes, invite them to order the cards correctly. For additional practice, you might even remove one or more cards when they are in the correct order and then ask students to identify the missing card(s).

Making and Using the Mini-Book

1. Share your pre-assembled mini-book with students. Read each page aloud and identify the tasks the students need to complete: Trace the dotted-line day of the week and record its ordinal position, 1st through 7th.
2. Distribute pages 77–80 and help students assemble according to the general directions on page 6.
3. Support students as they work through their mini-books. They may need help writing ordinal numbers. Then, invite students to decorate their mini-books.
4. Provide time for children to share their books, as a whole-class read-aloud, with partners, or with family members at home.

Taking It Further

Place students in groups of three or four and invite them to create their own alliterative day-of-the-week sentences. Assign each group a particular day. Then, have students work together to create a sentence with an accompanying illustration for that day. Students might create sentences such as *On Sunday we sing silly songs* or *On Monday we make magnificent mobiles,* and so on. Encourage the students to be creative!

The Mini-Books

I See Numbers

by ______________________

Juma

Two, two, two,
I see two.
Two tiny turtles
with nothing to do.

two

Add two triangle ▽ tails.

Juma

One, one, one,
I see one.
One lone octopus
having fun!

Add one oval balloon.

Juma

Three, three, three,
I see three.
Three old owls
up in a tree.

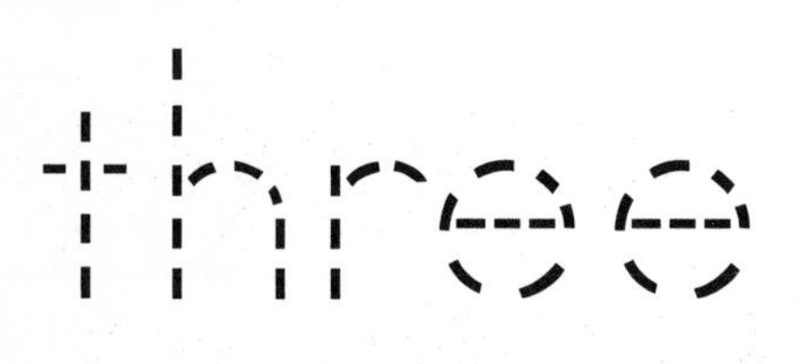

Add three triangle ▽ beaks.

Four, four, four,
I see four.
Four furry foxes
looking for more.

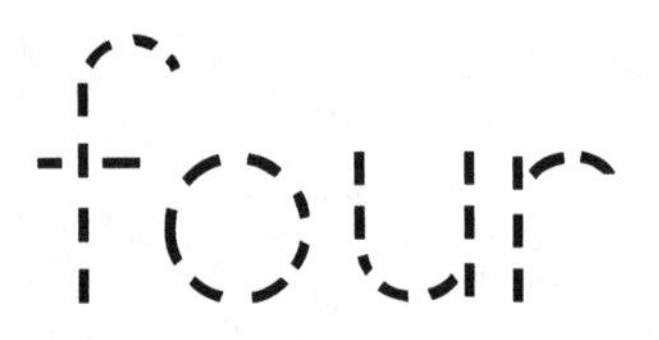

Add four circle ○ noses.

Six, six, six,
I see six.
Six silly snakes
playing some tricks.

Add six rectangle  tails.

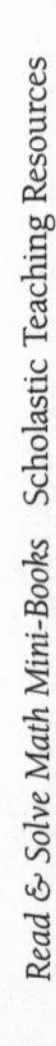

Juma

Five, five, five,
I see five.
Five funny fish
taking a dive.

Add five triangle ◁ fins.

Seven, seven, seven,
I see seven.
Seven slippery snails
slithering to Devon.

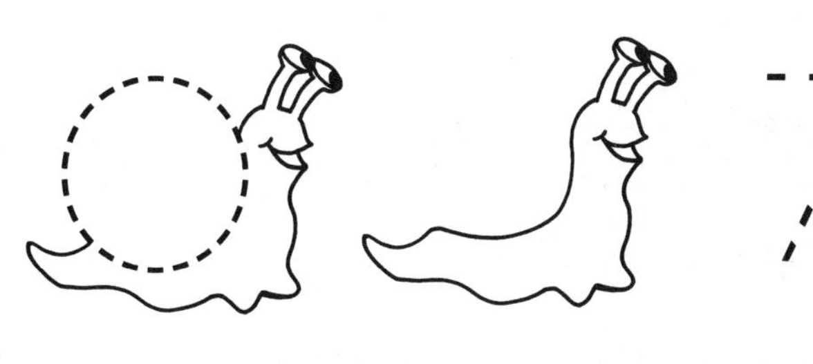

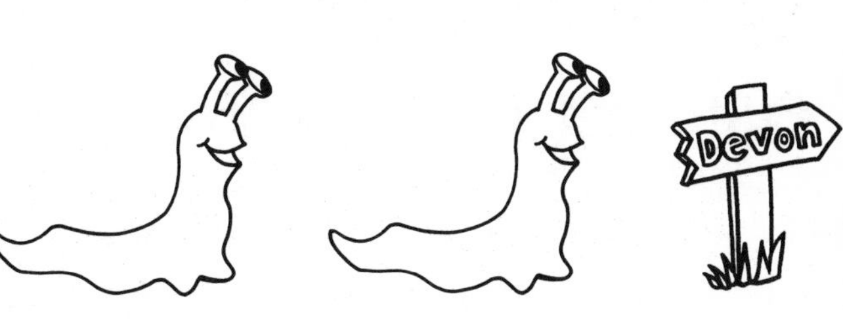

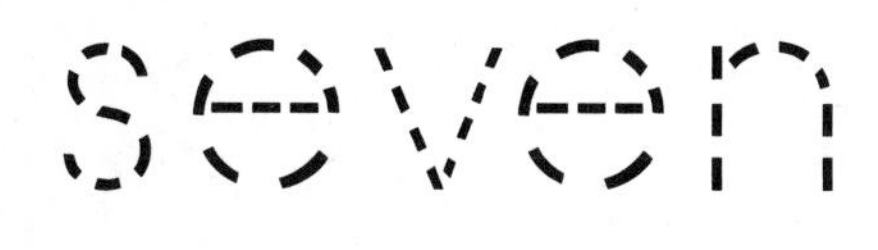

Add seven circle ◯ shells.

Eight, eight, eight,
I see eight.
Eight enormous elephants
who all have to wait.

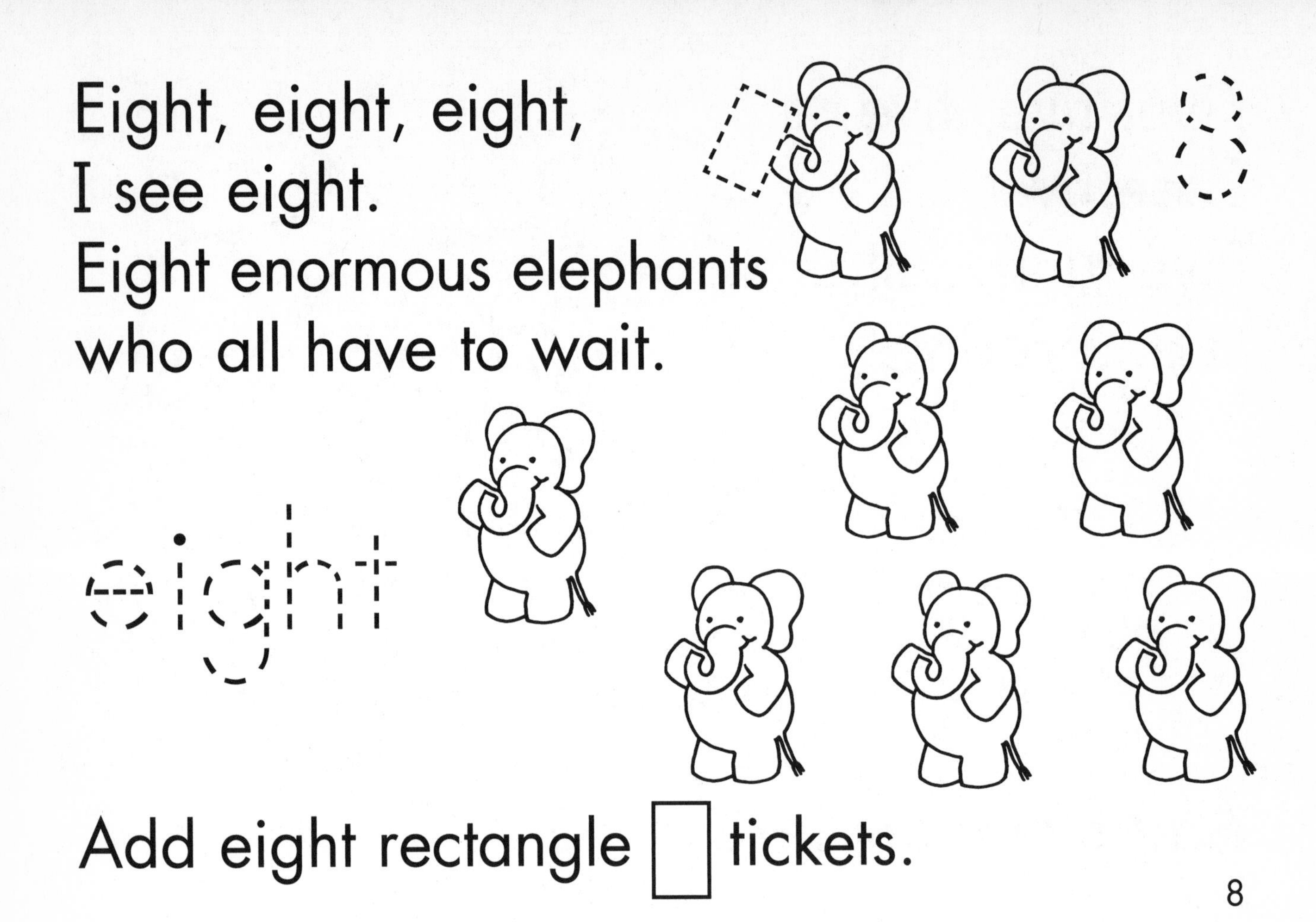

Add eight rectangle ▭ tickets.

Ten, ten, ten,
I see ten.
Ten tired tigers
asleep in a den.

Add ten circle ○ mouths.

Nine, nine, nine,
I see nine.

Nine noisy nightingales,
all out to dine.

Add nine square □ breadcrumbs.

Numbers, numbers,
numbers,
I see numbers.
One to ten,
count them again!

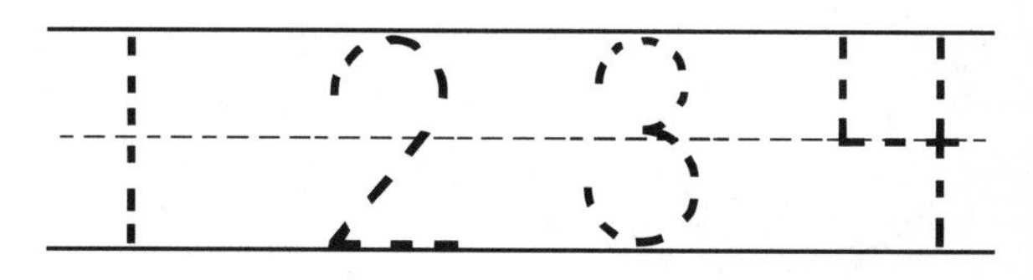

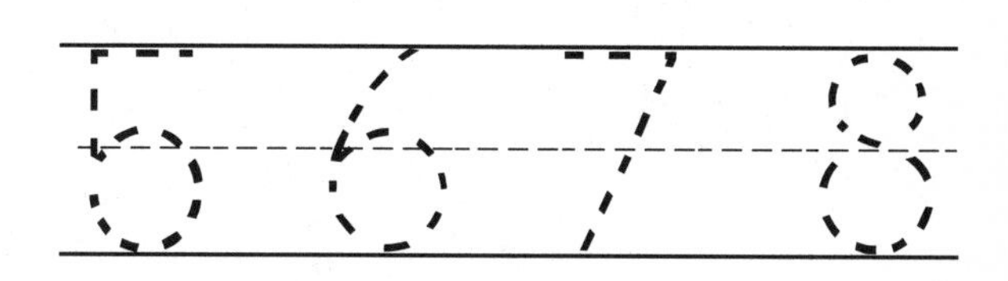

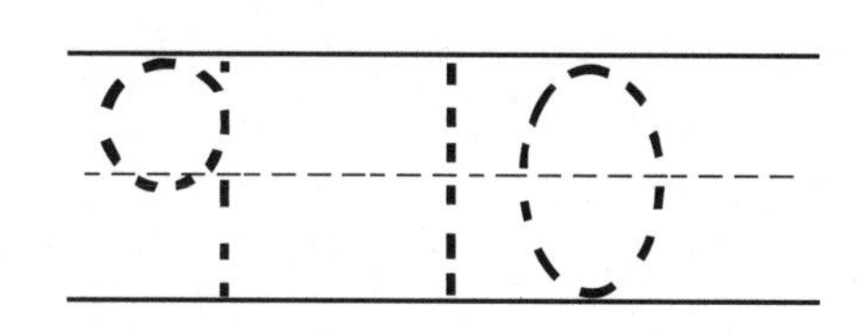

Missing Numbers!

by ______________________________

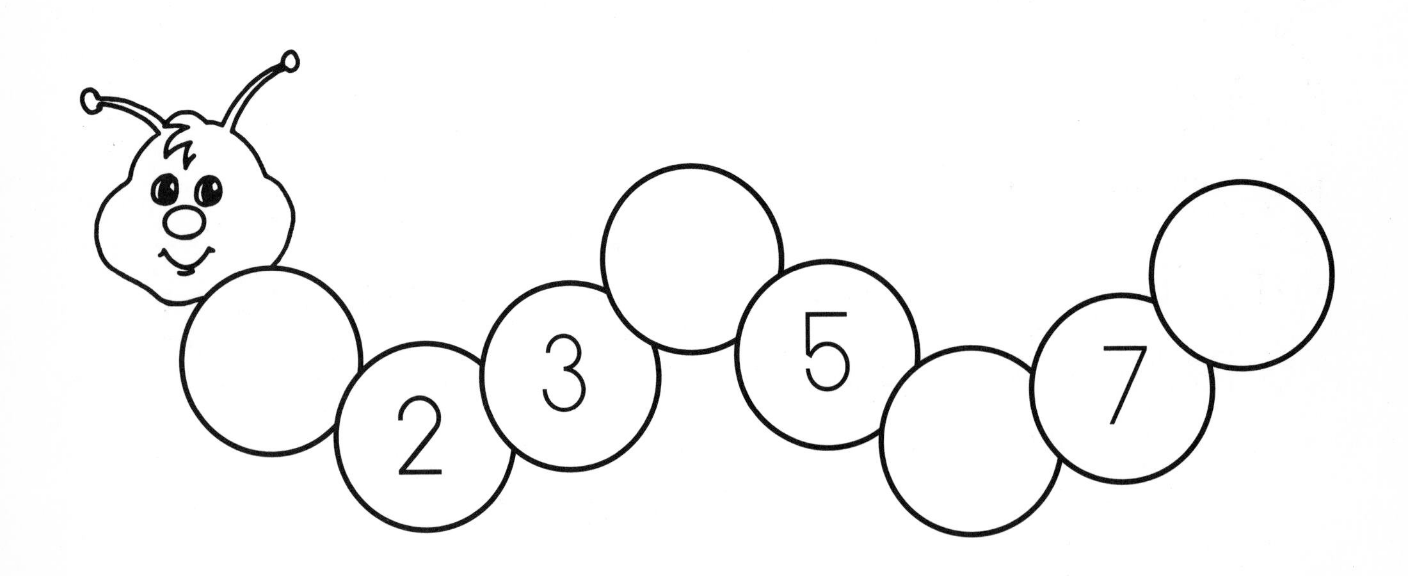

This medium caterpillar is missing numbers. Can you fill them in?

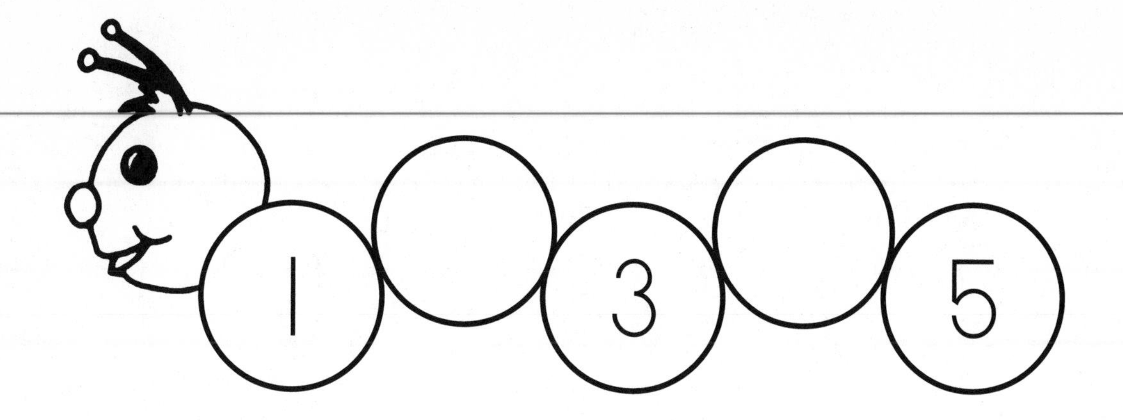

This short caterpillar is missing numbers.
Can you fill them in?

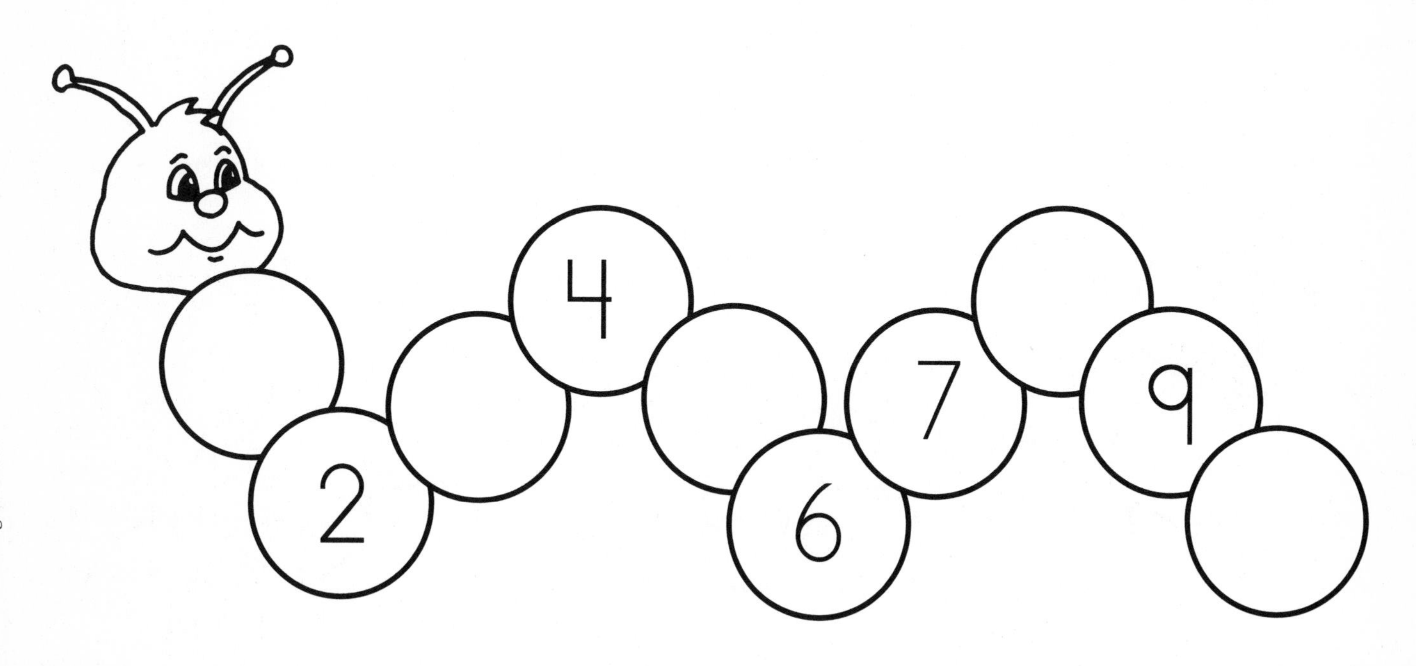

This long caterpillar is missing numbers.
Can you fill them in?

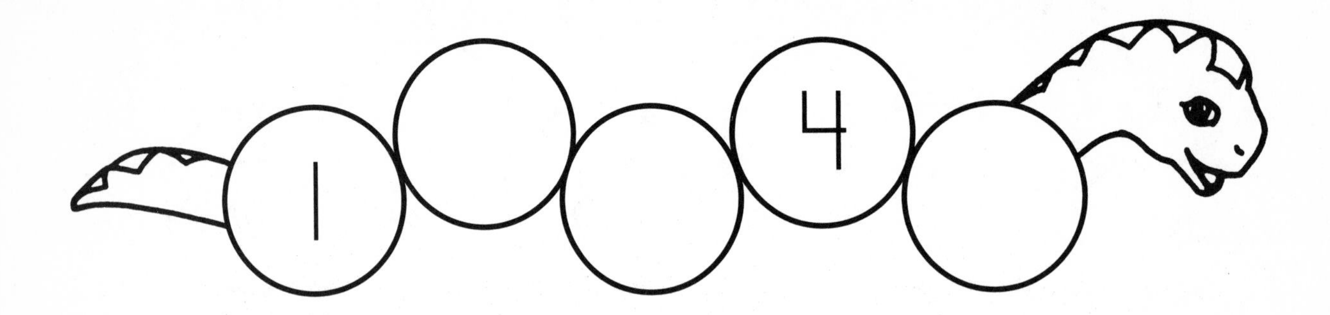

This short snake is missing numbers.
Can you fill them in?

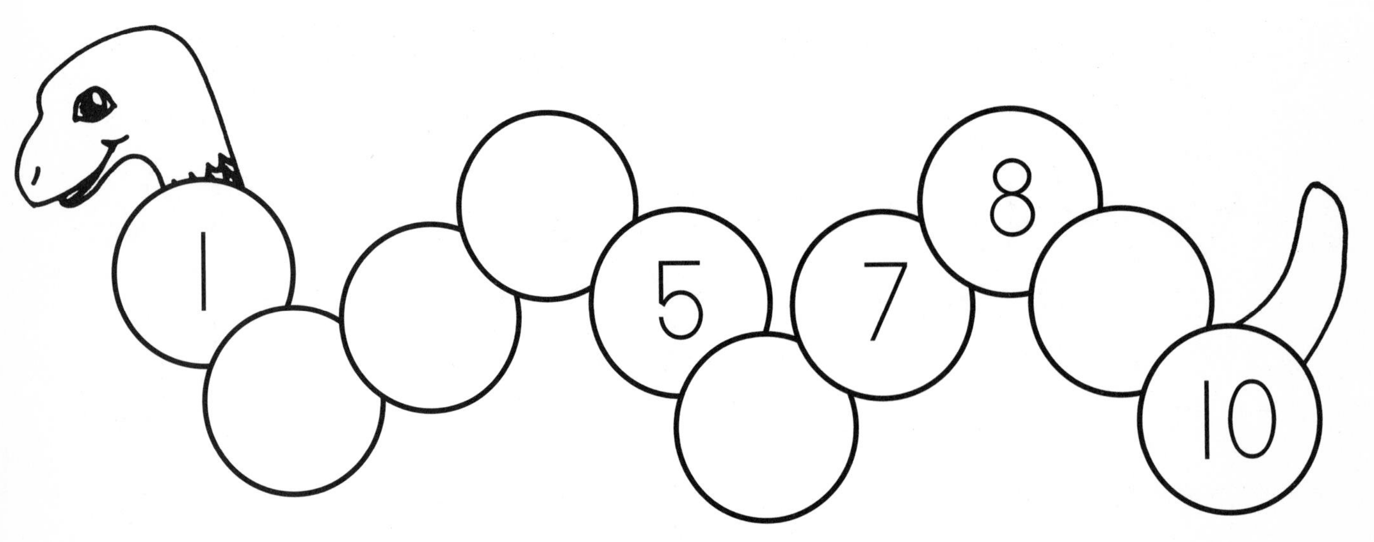

This long snake is missing numbers.
Can you fill them in?

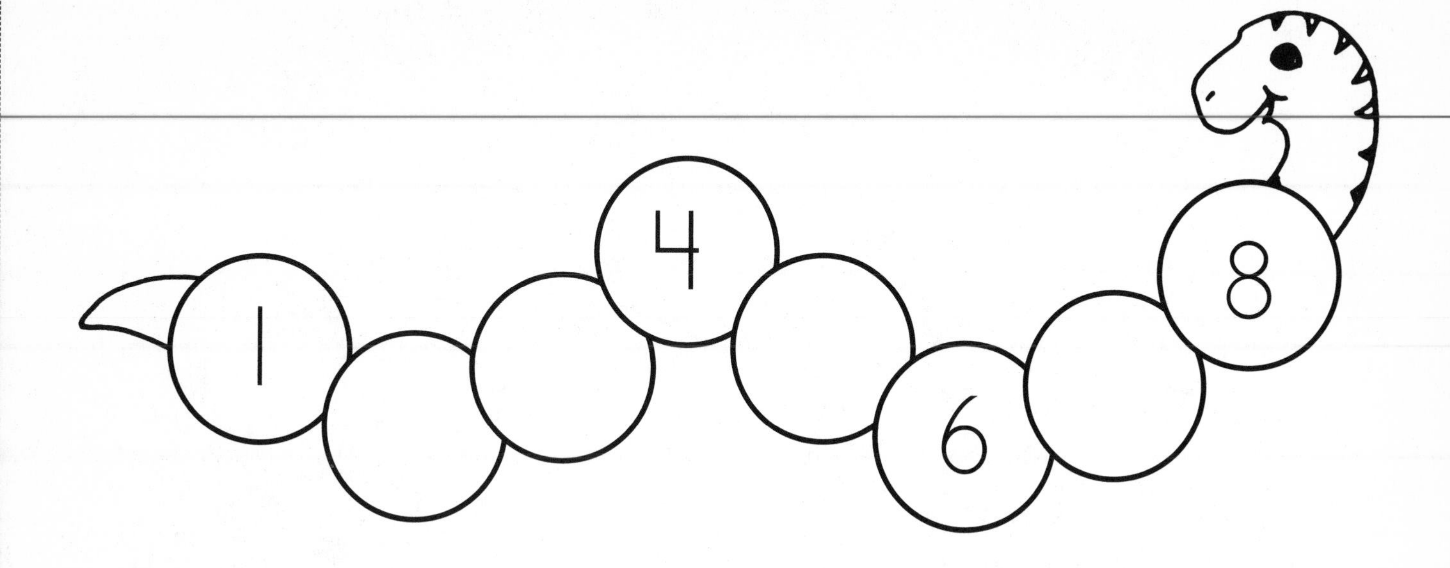

This medium snake is missing numbers.
Can you fill them in?

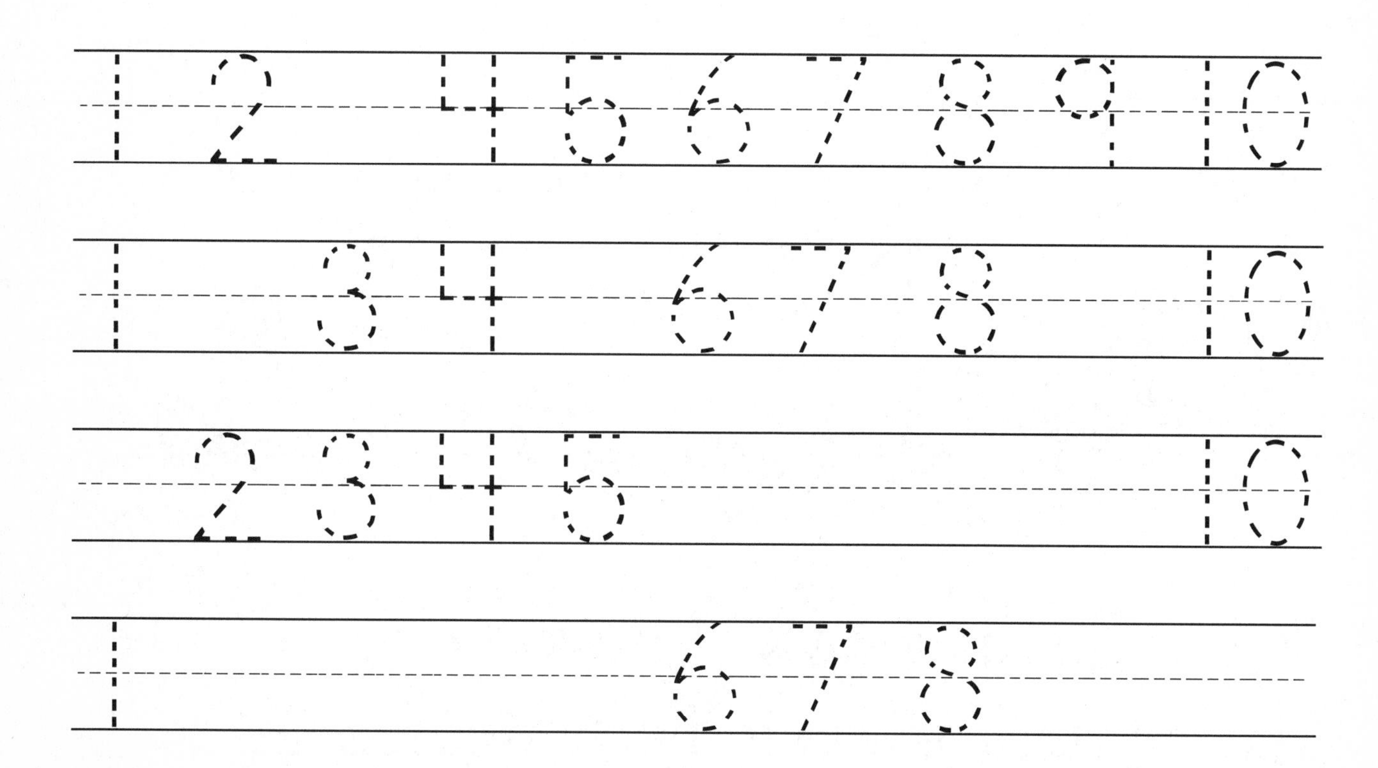

Print the numbers one to ten!

One More Apple

by ______________________________

My basket is empty. I have none.
Into my basket I put the first ________!

____ + ____ = ____

Wishing for apples, and what do I see?
Lots of shiny red apples up in a tree!

1

I have one apple just for you.
I pick one more and that makes ________.

____ + ____ = ____

I find one more under a tree.
All together that makes ____________.

____ + ____ = ____

Number ____________ is next you know.
Count them fast or count them slow.

____ + ____ = ____

Then I find one apple more.
All together I have ____________.

____ + ____ = ____

5

I sort my apples. I have a mix.
Five red plus one green equal ____________.

____ + ____ = ____

7

I have ____________ when I add one more.
I can stack them on the floor.

____ + ____ = ____

A bushel of eight apples, isn't it fine?
I add one apple and now I have ________.

____ + ____ = ____

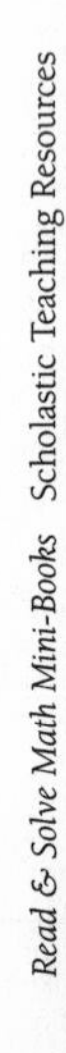

Apples in a row are nice and straight.
Add one to seven and I have __________.

____ + ____ = ____

I add one more apple. Do you know why?
__________ is enough to make a pie!

____ + ____ = ____

How Many in All?

by ______________________

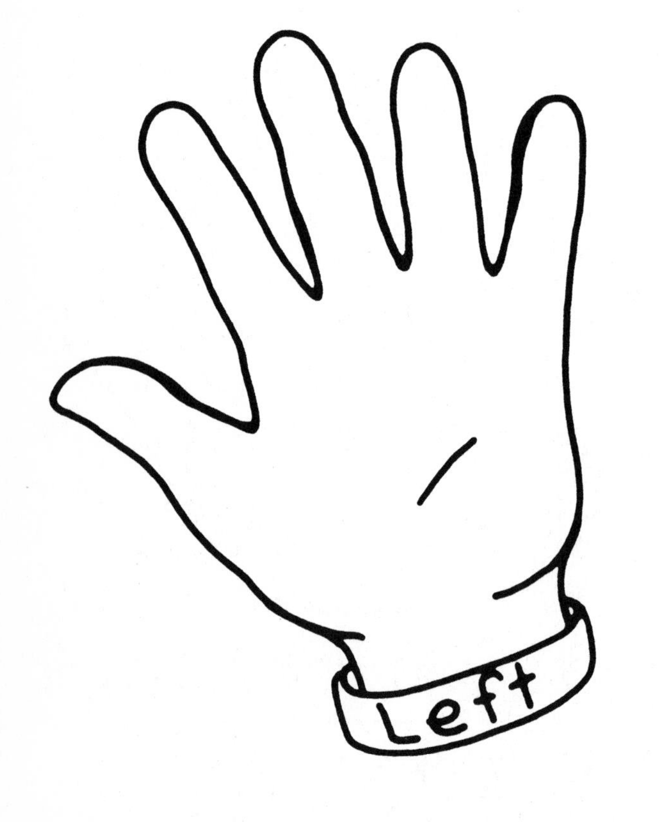

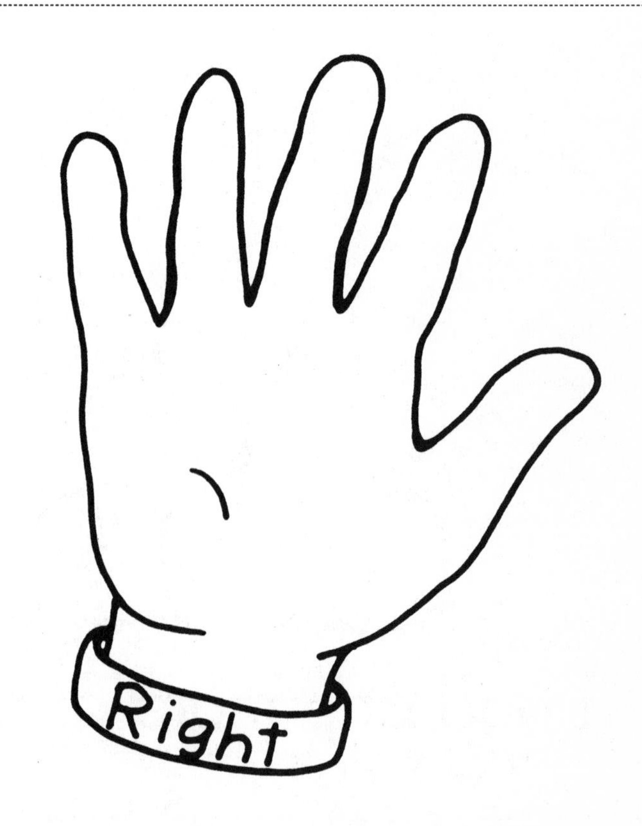

$1 + 2 =$ ____

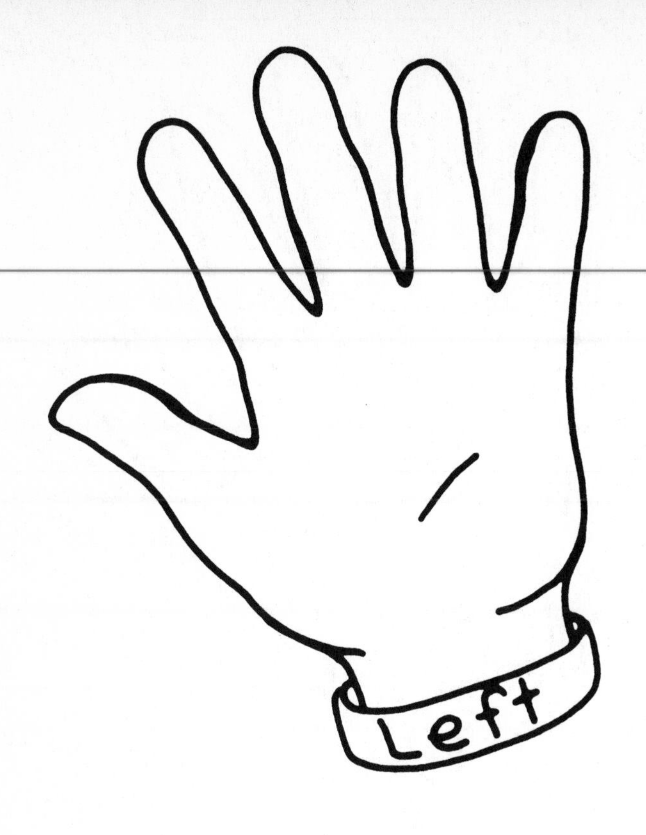

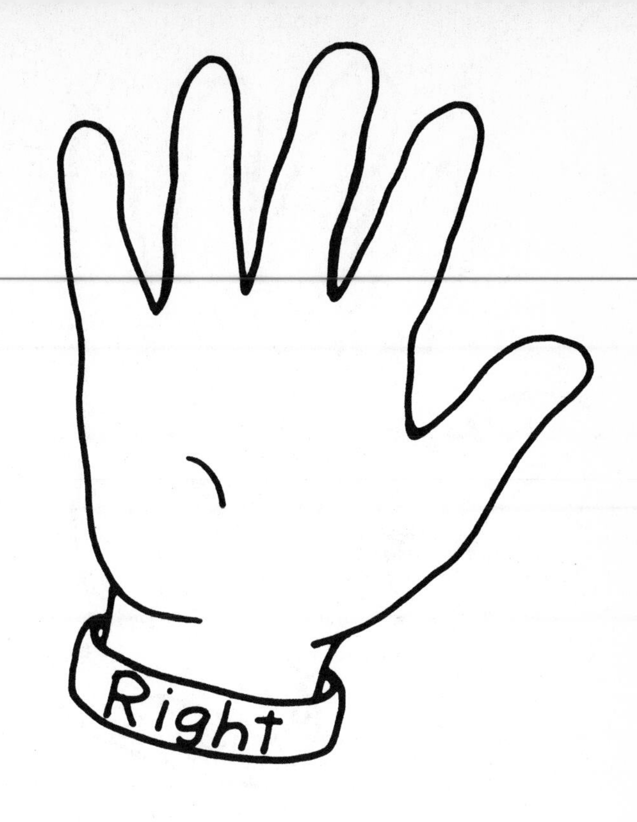

1 + 1 = ____

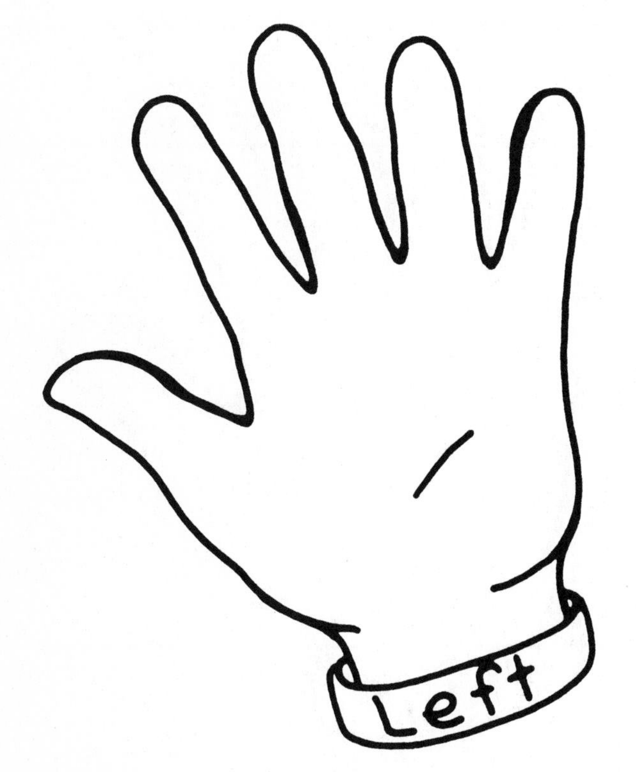

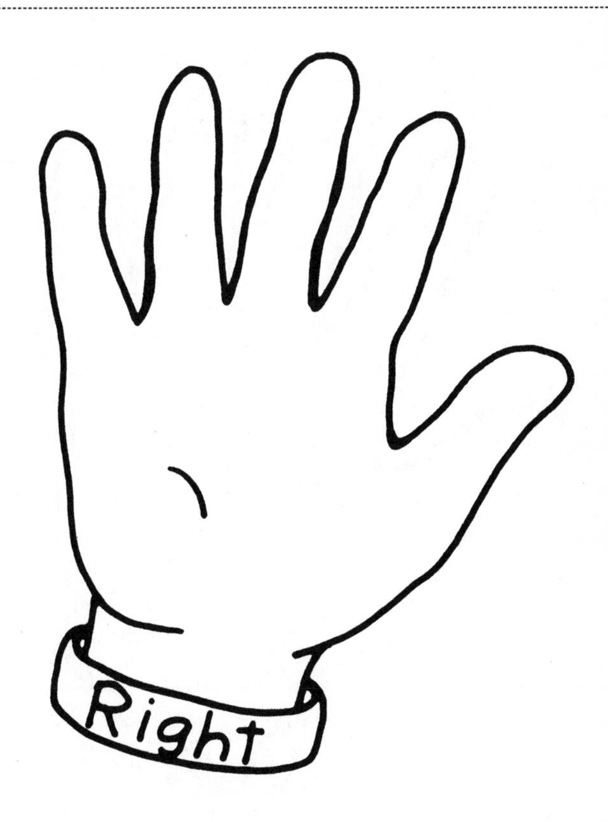

0 + 3 = ____

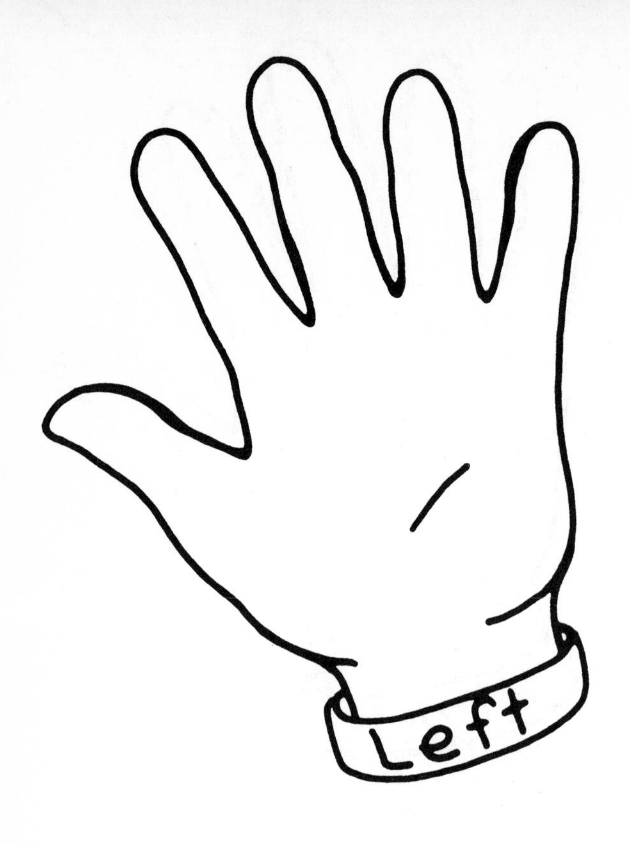

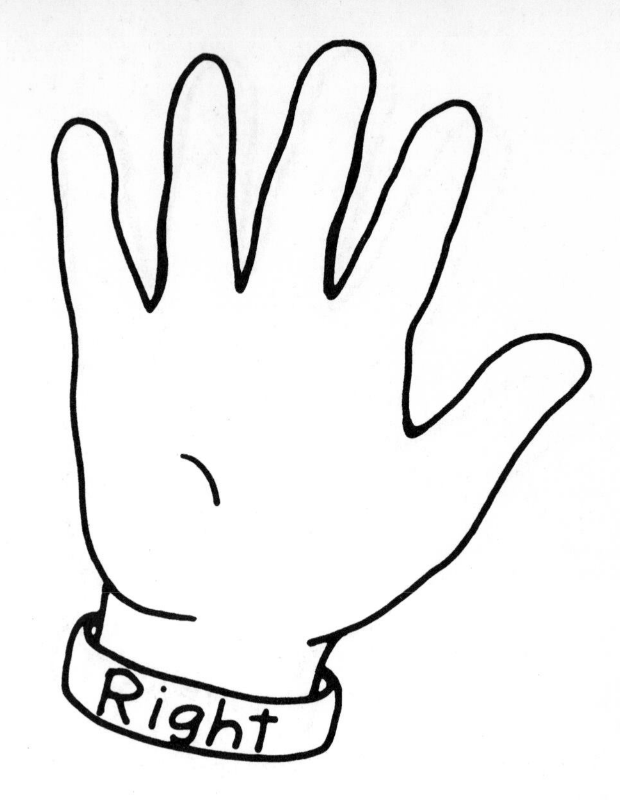

2 + 2 = ____

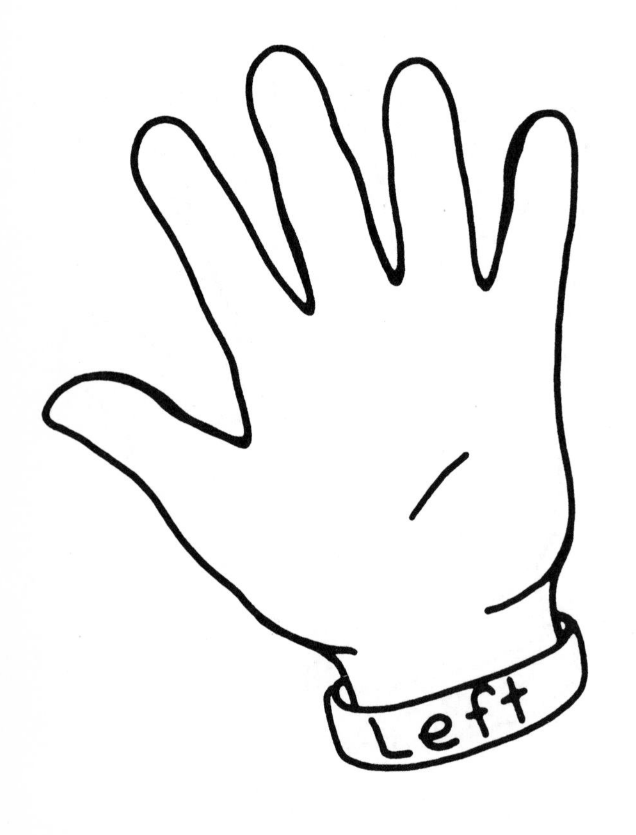

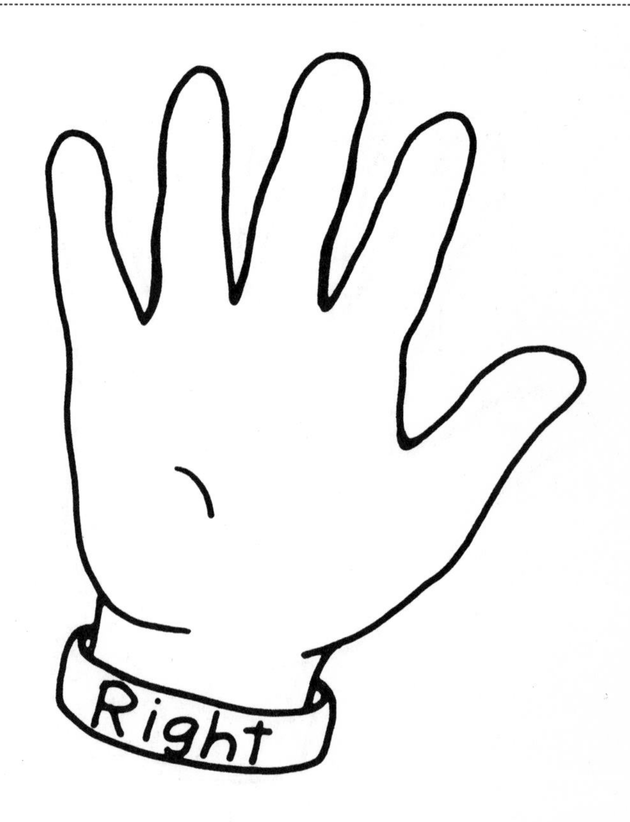

2 + 3 = ____

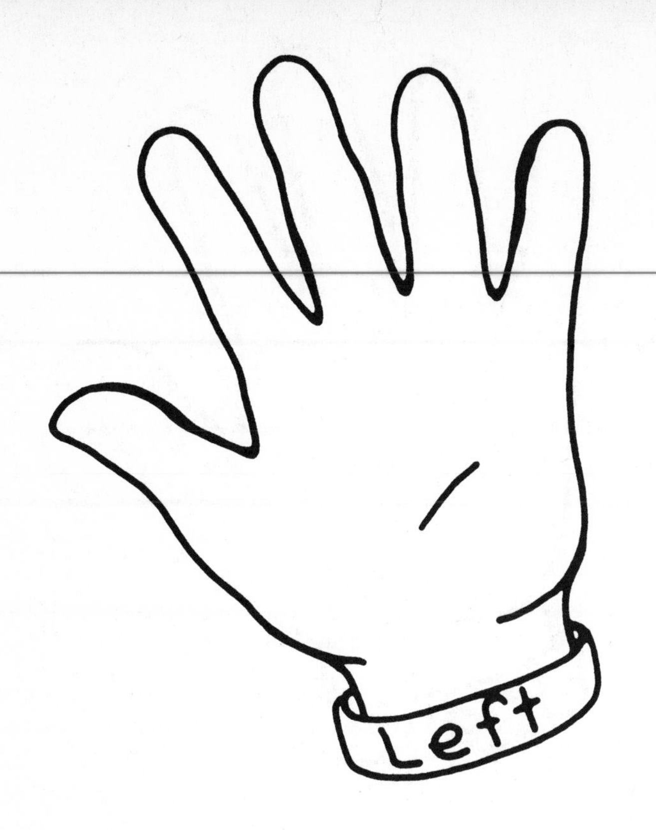

$3 + 1 =$ _____

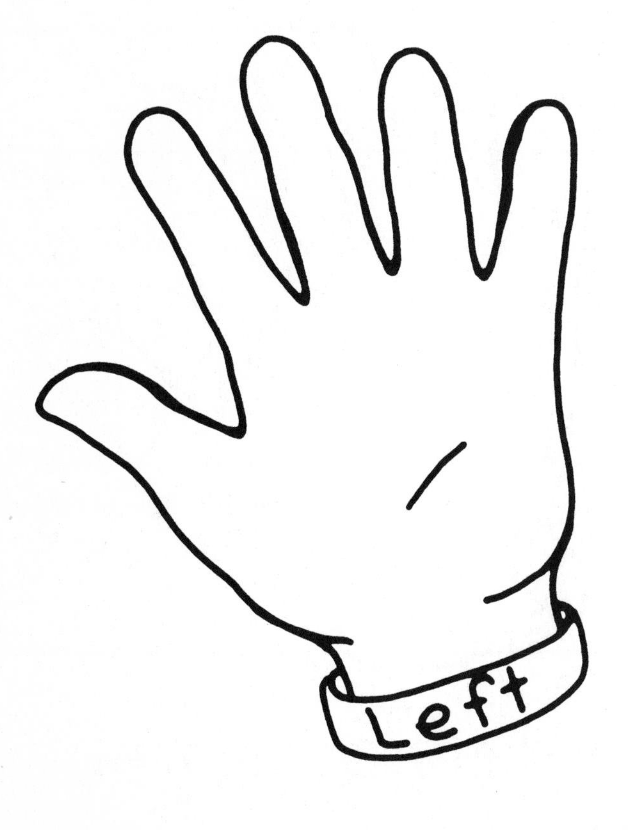

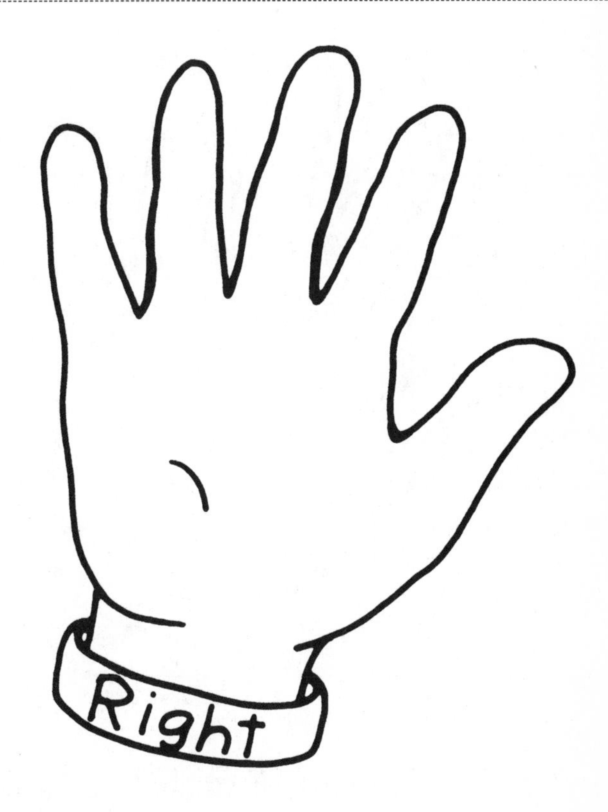

$4 + 0 =$ _____

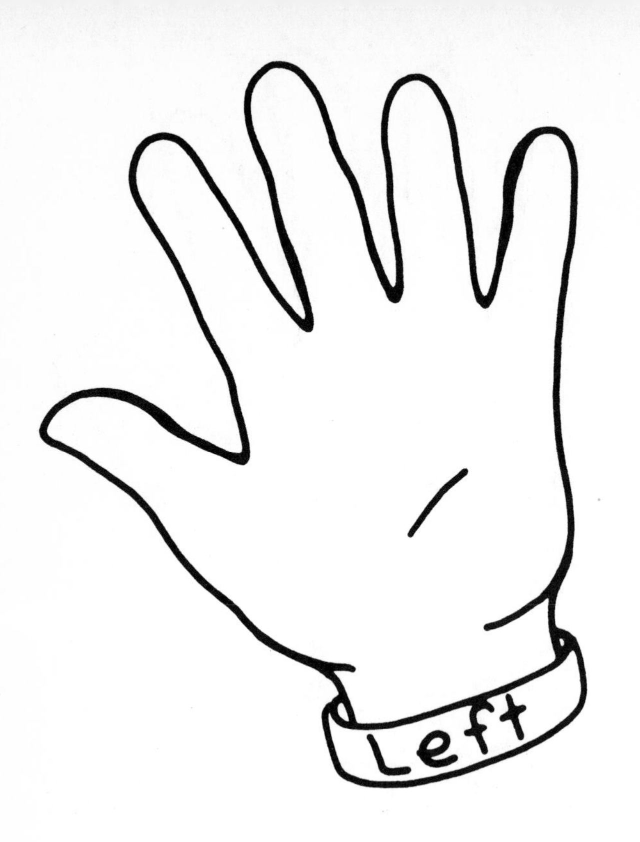

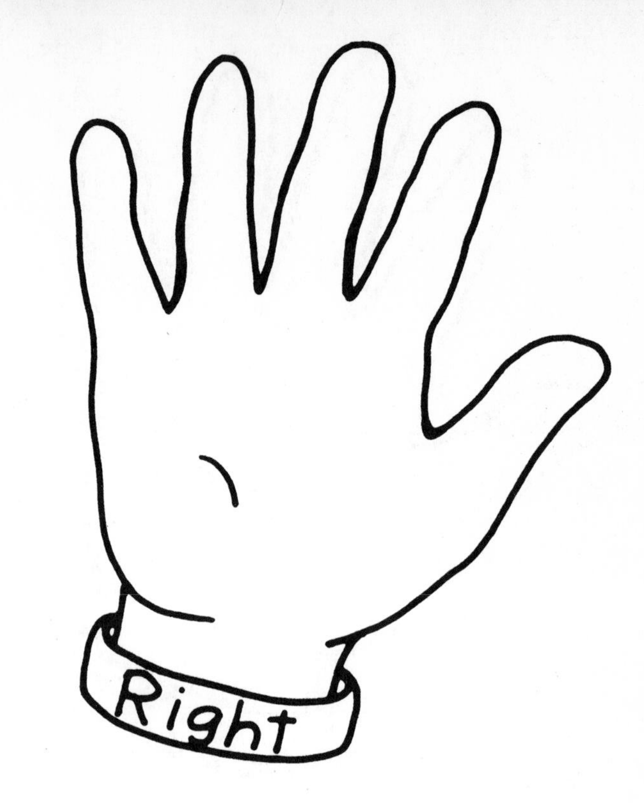

$1 + 5 =$ ____

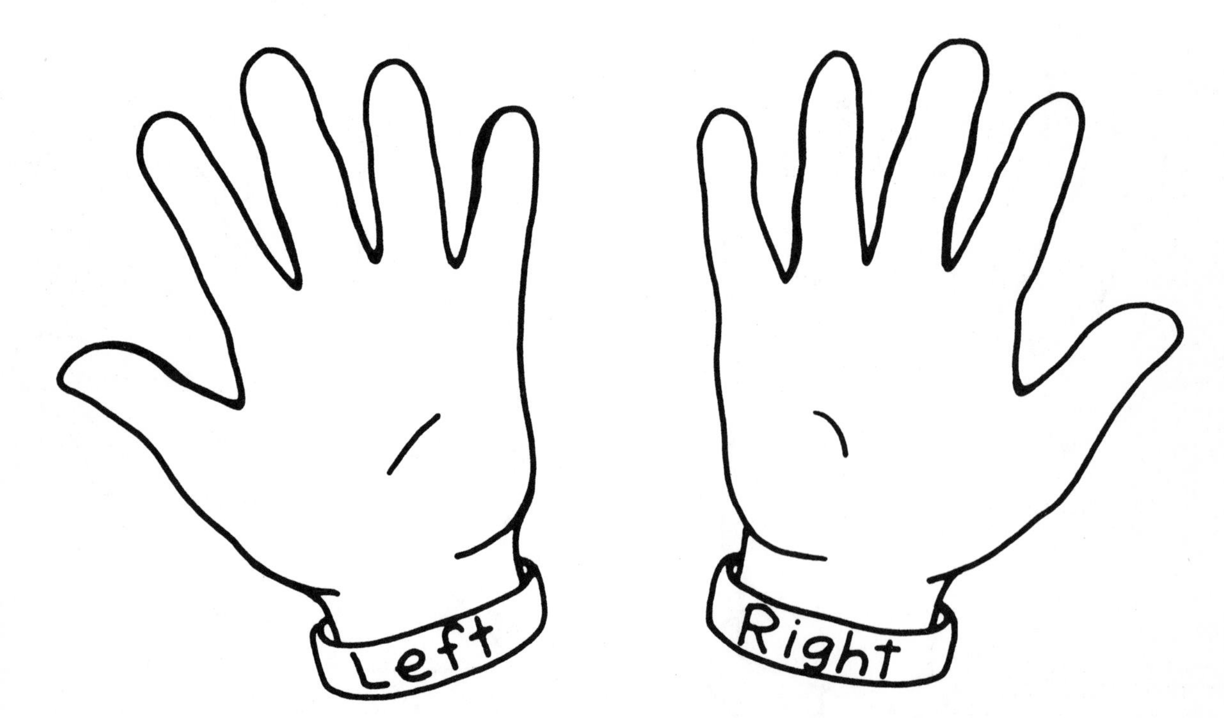

I can make my own addition sentence!

____ + ____ = ____

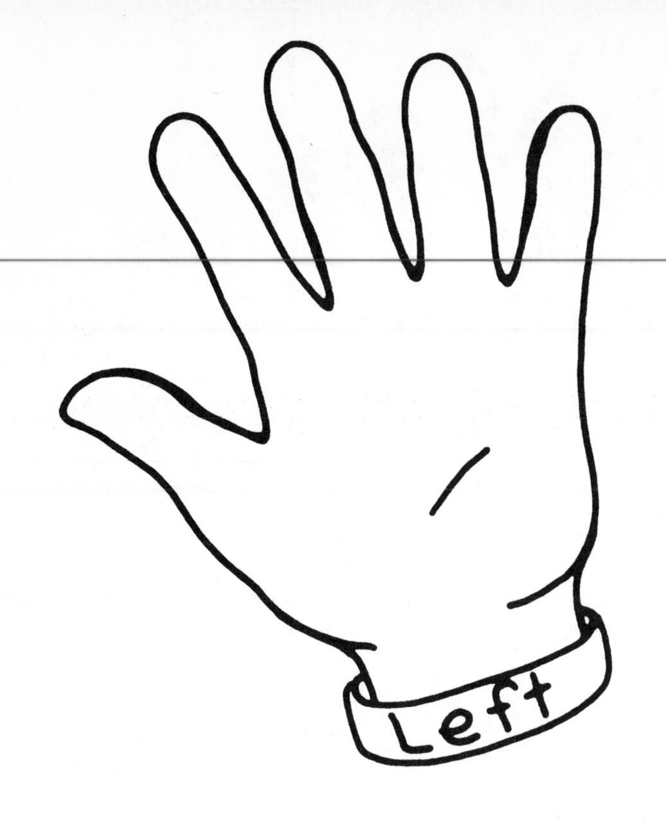

$4 + 3 =$ ____

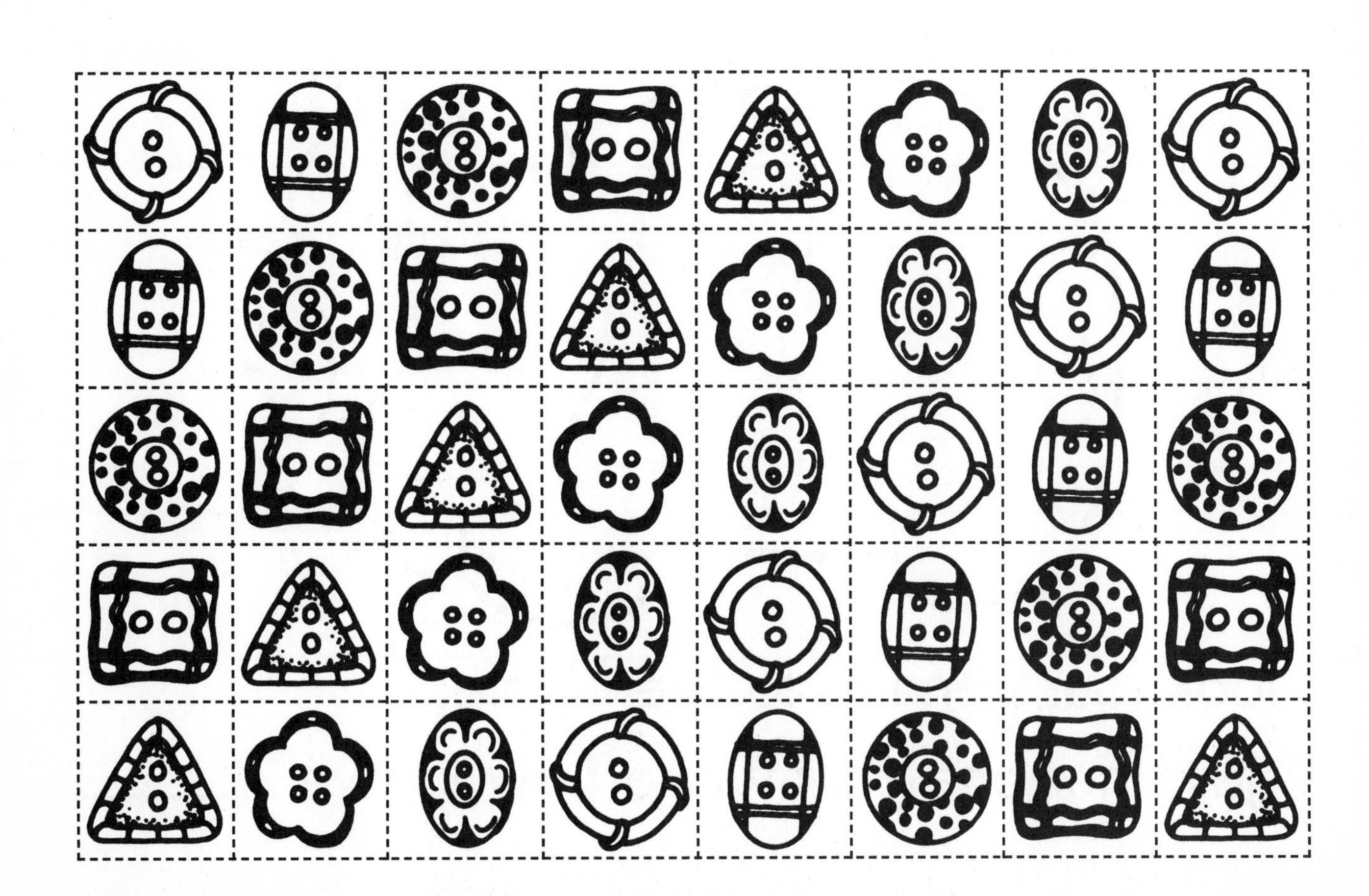

Five Pennies to Spend

by ______________________________

I took four pennies to see Marie.

I gave her one penny and now I have ______.

____ – ____ = ____

I took five pennies to the store.

I spent one penny and now I have ________.

____ – ____ = ____

1

I took three pennies to the zoo.

I spent one penny and now I have ________.

____ – ____ = ____

I took two pennies out in the sun.
I left one penny and now I have __________.

____ – ____ = ____

I went home with zero pennies,
but then I did a chore.
Mom said, "Great job!" and gave me five more.

____ + ____ = ____

I took one penny to see a show.

I spent that penny and now I have ________.

___ – ___ = ___

Shapes All Around

by ______________________

Juma

____

square

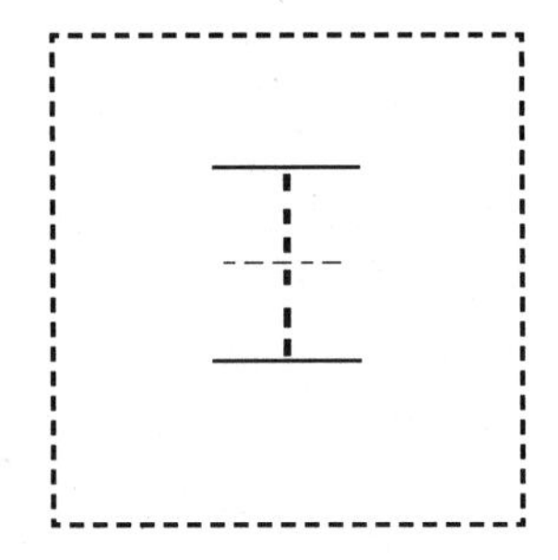

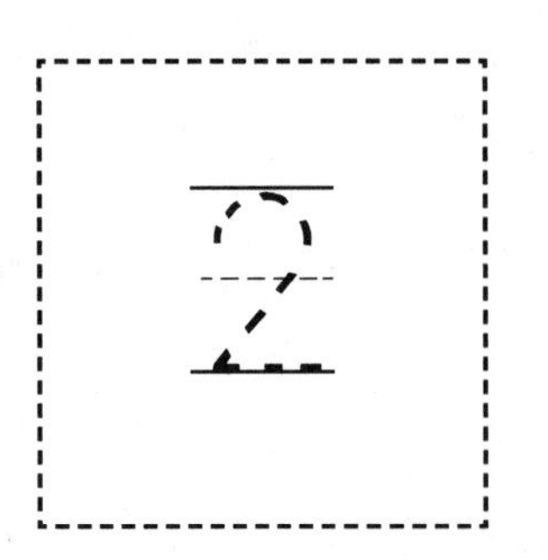

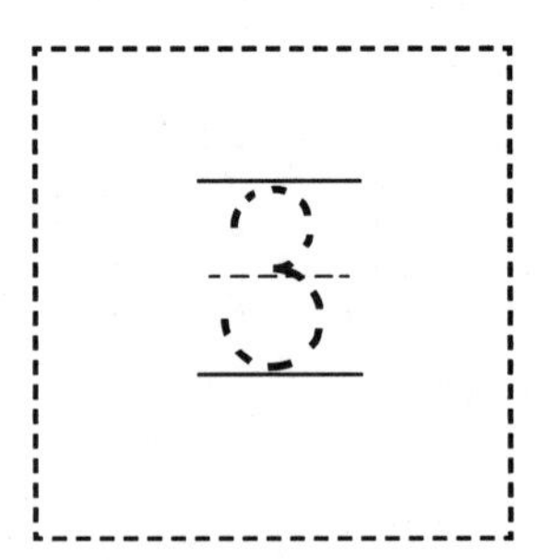

Trace the word square black.
Trace the squares ⬚ red.
Trace the numbers yellow.
How many squares in all? _____

circle

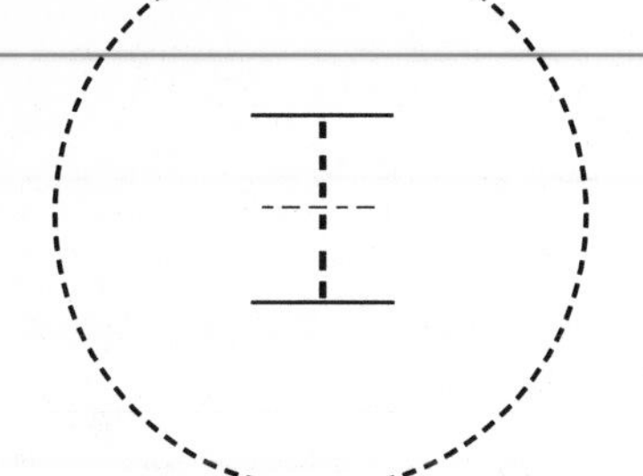

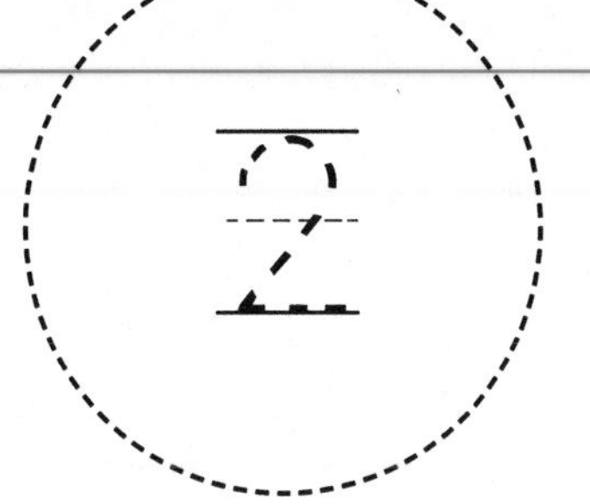

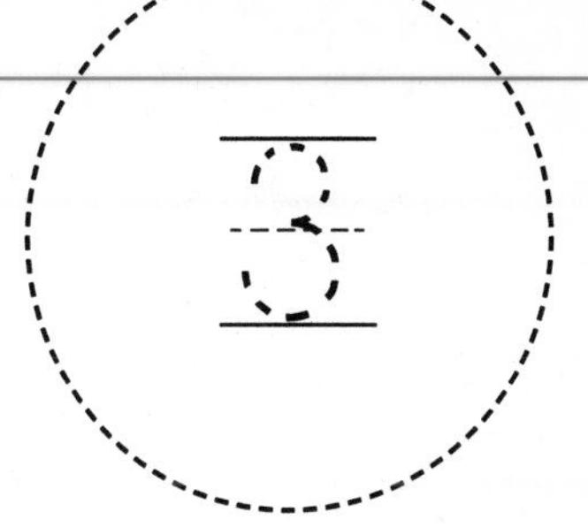

Trace the word circle black.
Trace the circles red.
Trace the numbers yellow.
How many circles in all? _____

rectangle

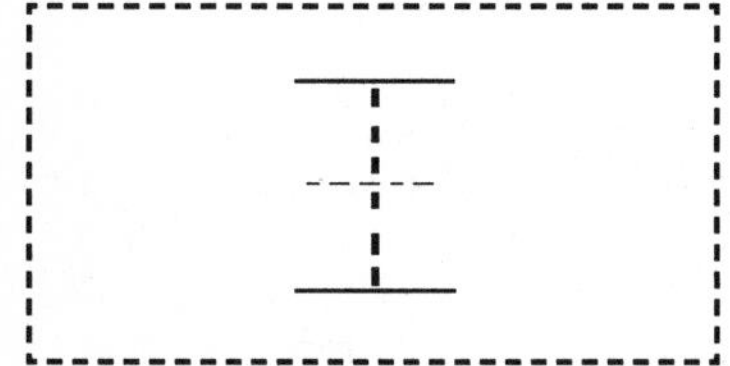

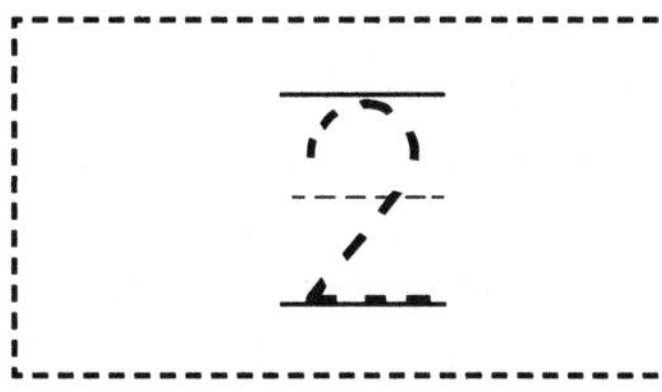

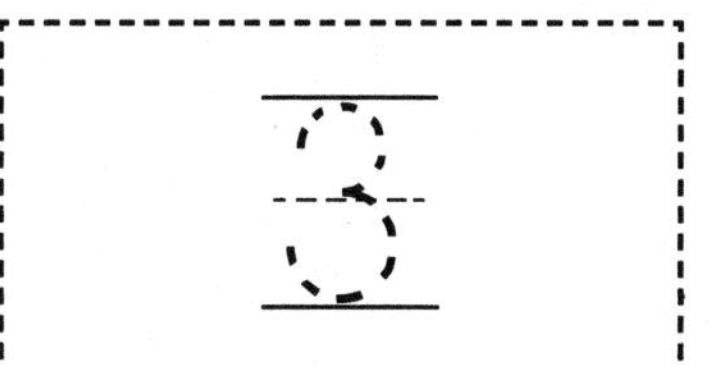

Trace the word rectangle blue.
Trace the rectangles green.
Trace the numbers orange.
How many rectangles in all? _____

Juma

triangle

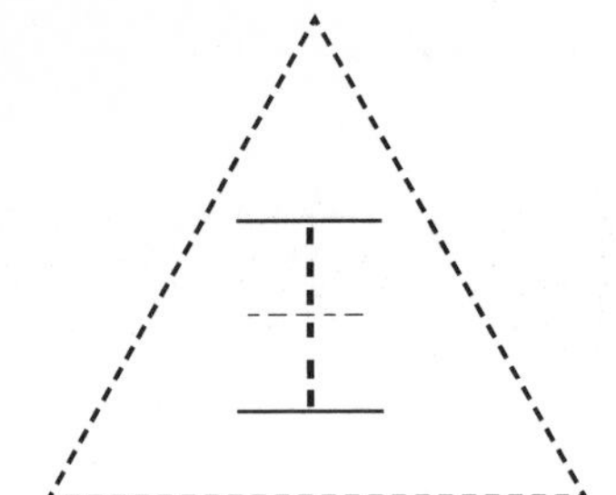

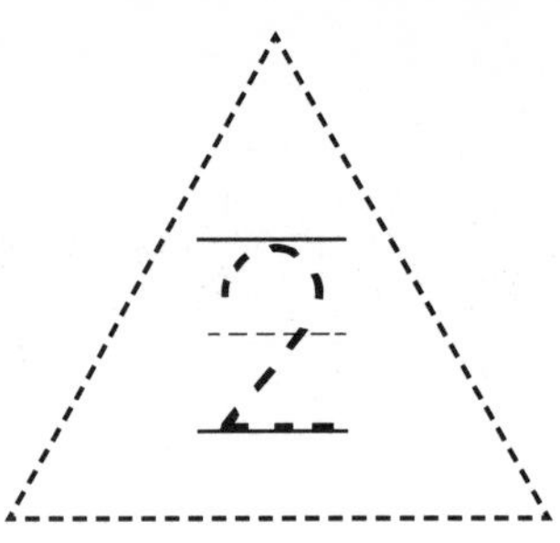

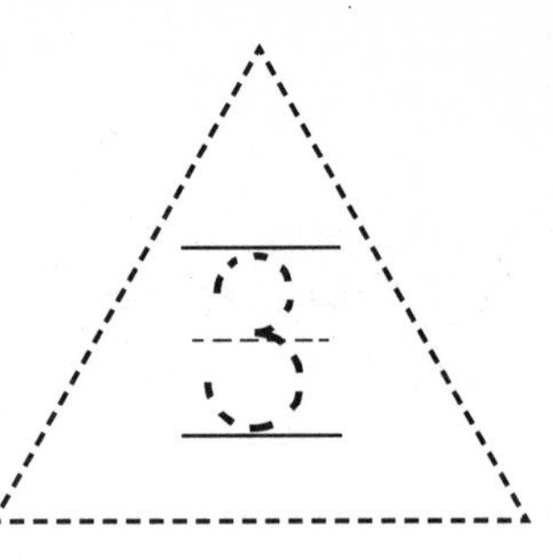

Trace the word triangle blue.

Trace the triangles green.

Trace the numbers orange.

How many triangles in all? _____

How many squares in all? _____

How many triangles in all? _____

How many rectangles in all? _____

How many circles in all? _____

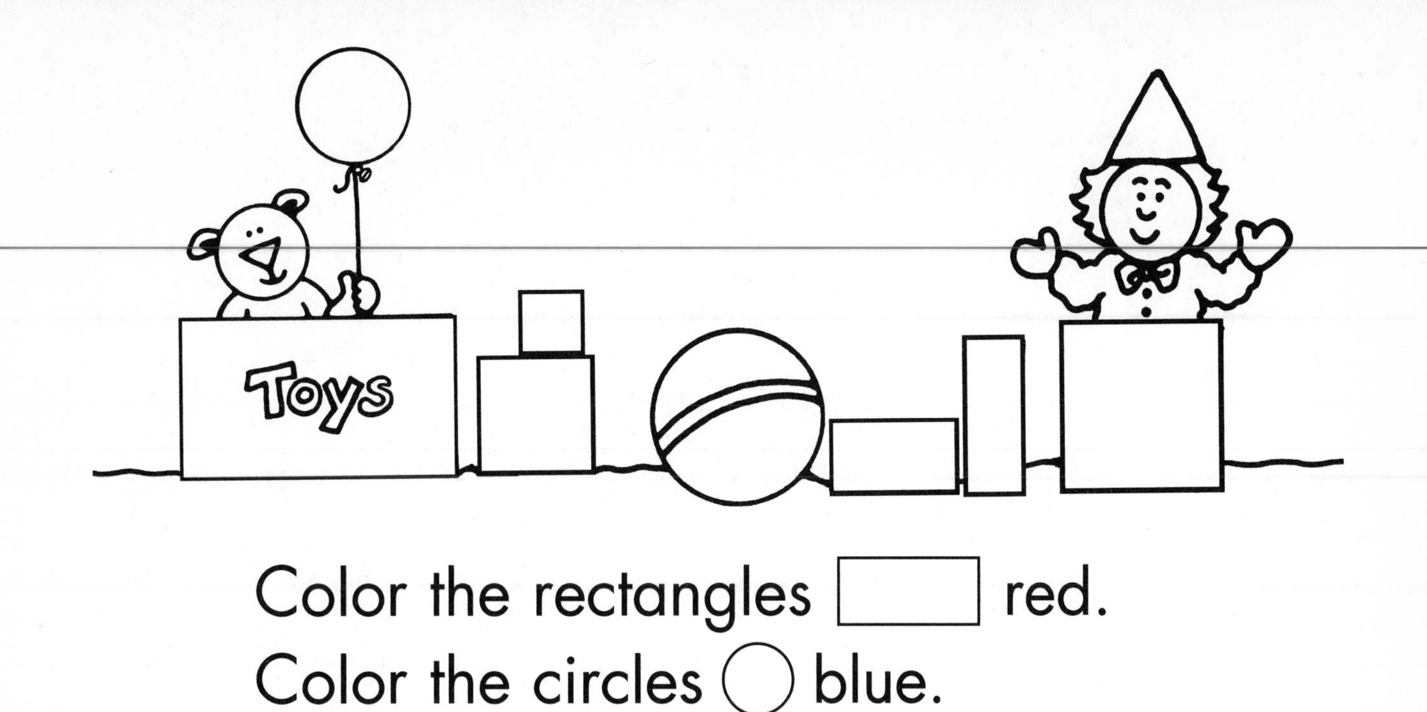

Color the rectangles ▭ red.
Color the circles ○ blue.
Color the triangles △ yellow.
Color the squares □ green.

Draw two of each shape!

Super Shapes!

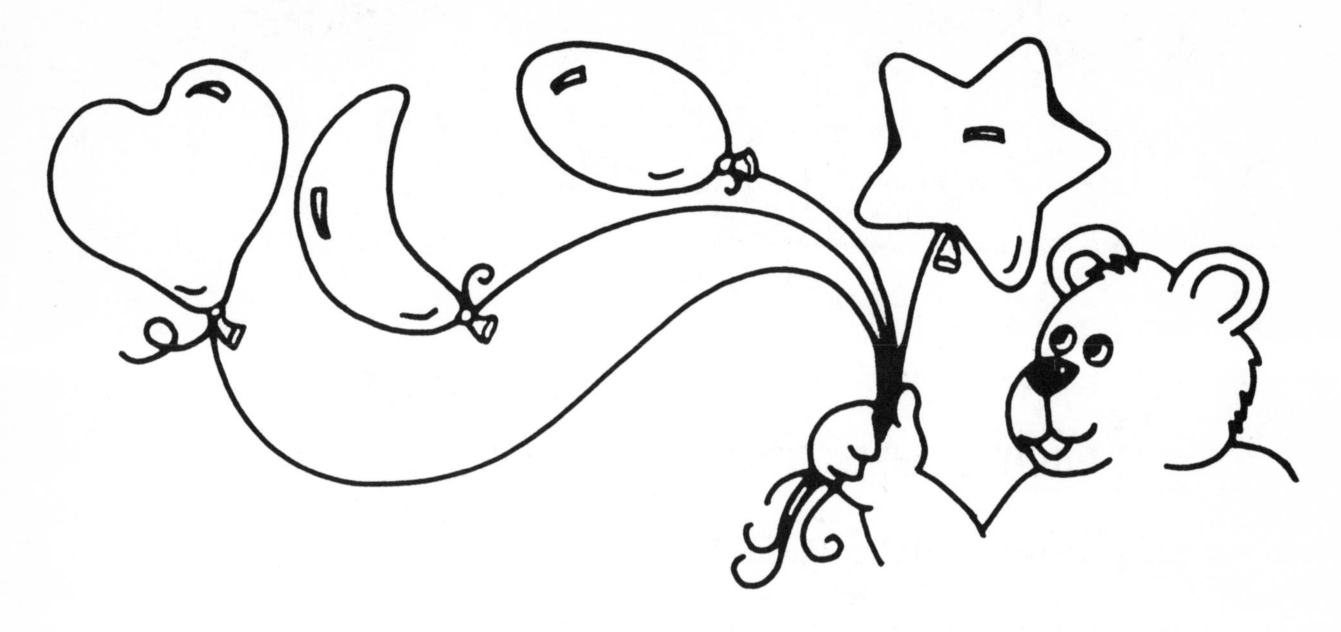

by ______________________________

Juma

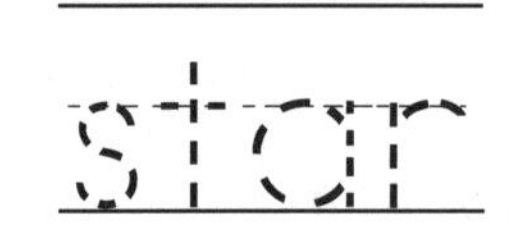

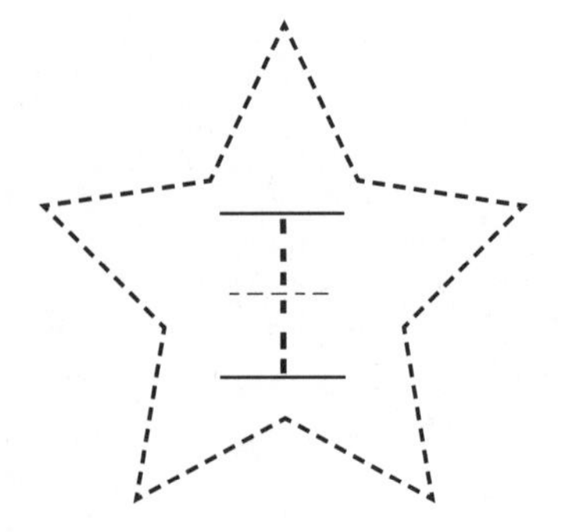

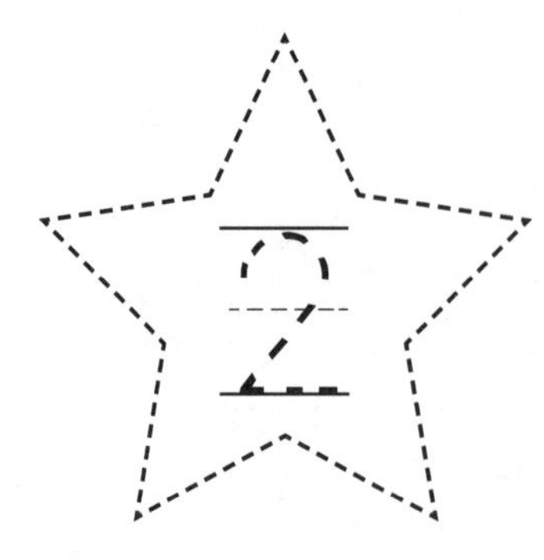

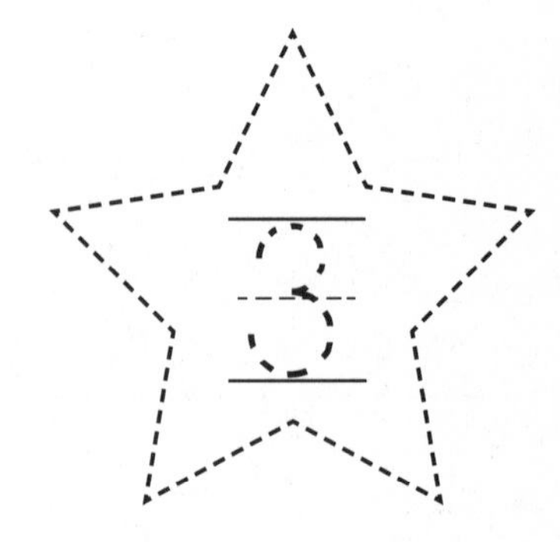

Trace the word star blue.
Trace the stars ☆ green.
Trace the numbers red.
How many stars in all? _____

Juma

oval

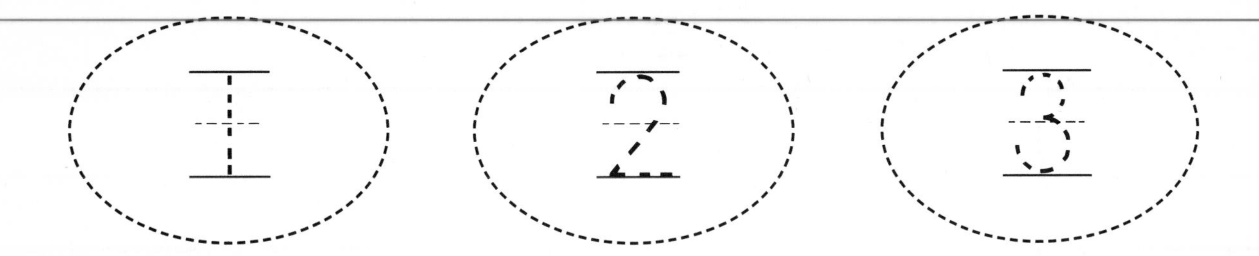

Trace the word oval blue.
Trace the ovals green.
Trace the numbers red.
How many ovals in all? ____

Juma

crescent

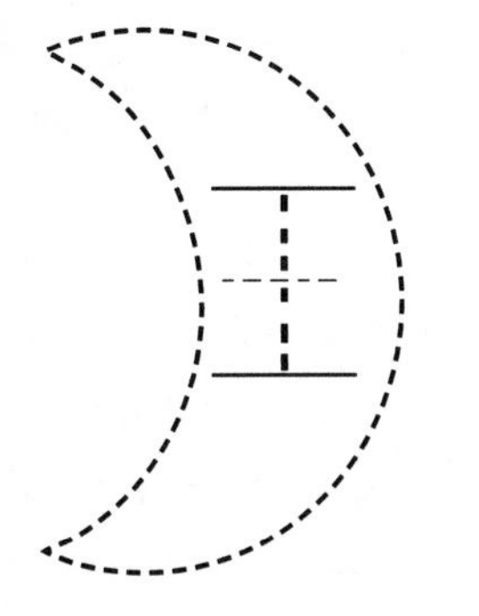

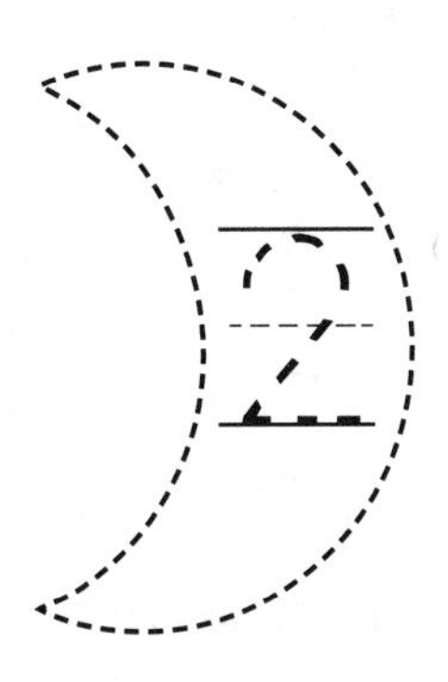

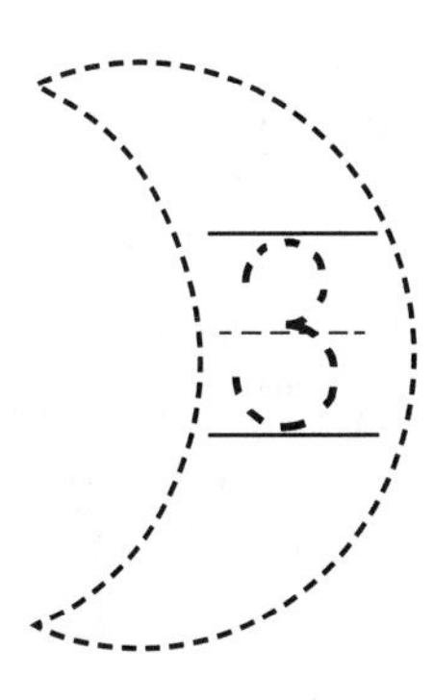

Trace the word crescent blue.
Trace the crescents orange.
Trace the numbers yellow.
How many crescents in all? ____

Juma

heart

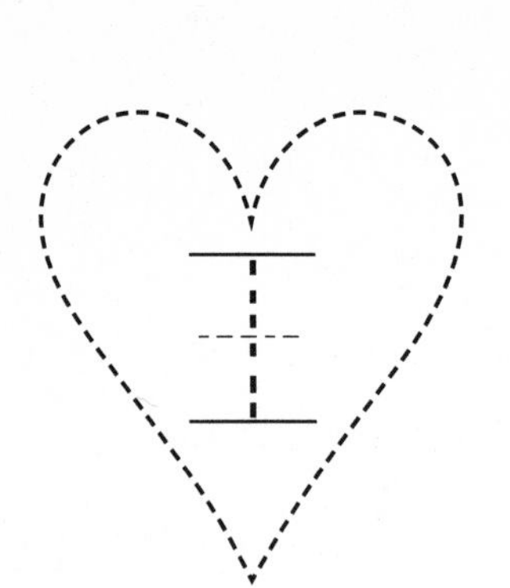
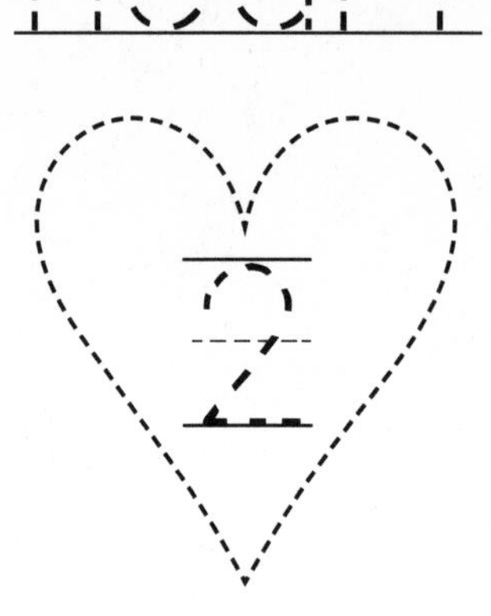
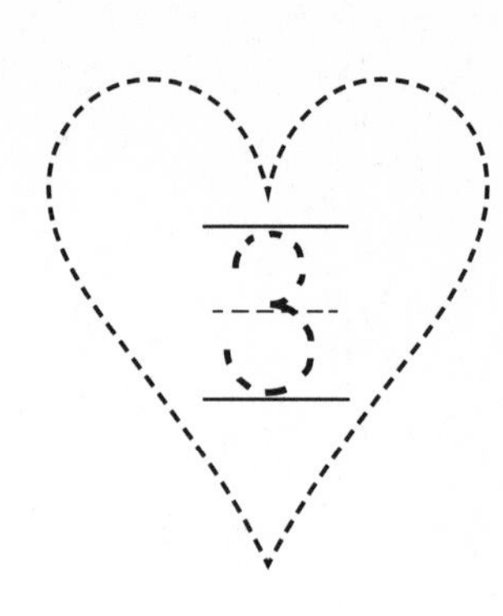

Trace the word heart blue.
Trace the hearts ♡ orange.
Trace the numbers yellow.
How many hearts in all? _____

How many hearts ♡ in all? _____
How many ovals ⬭ in all? _____
How many crescents ☽ in all? _____
How many stars ☆ in all? _____

Color the crescents green.
Color the stars yellow.
Color the hearts red.
Color the ovals orange.

Draw two of each shape!

Patterns Everywhere

by ______________________________

A	B	A	B	A	B

A	B	A	B	___	___

Extend the pattern!

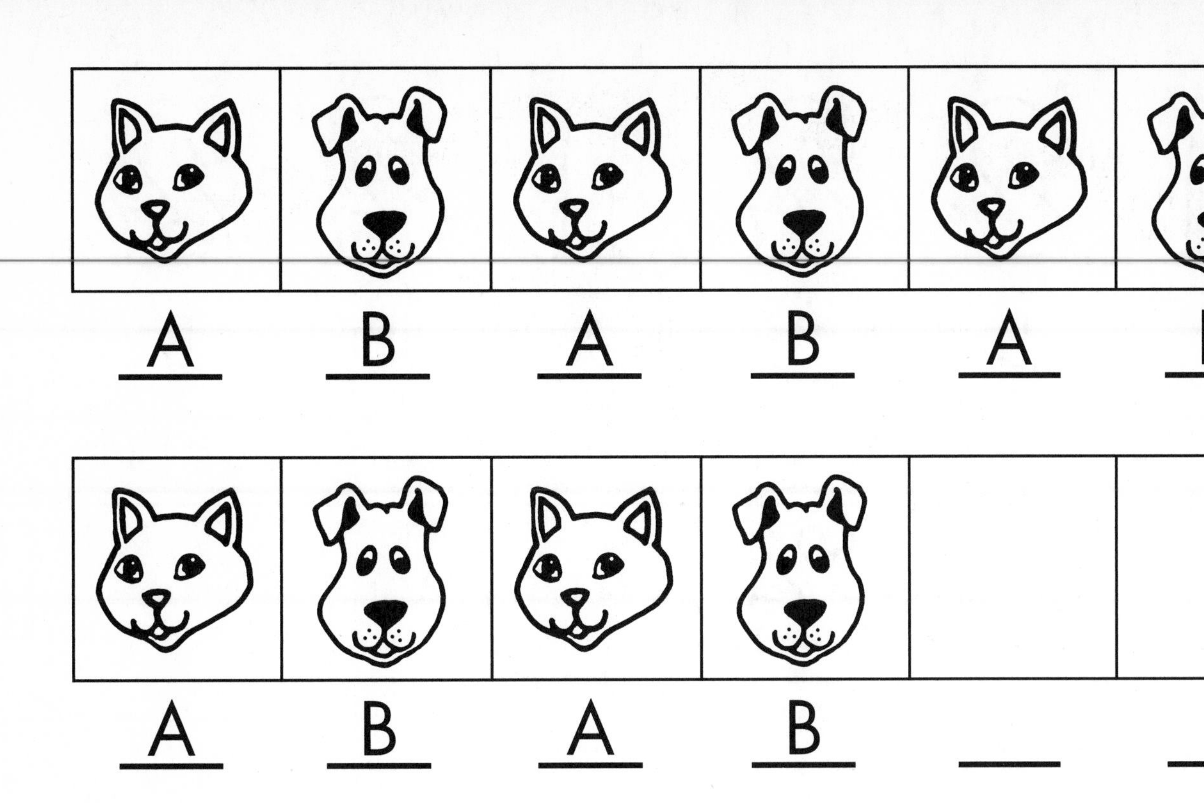

Extend the pattern!

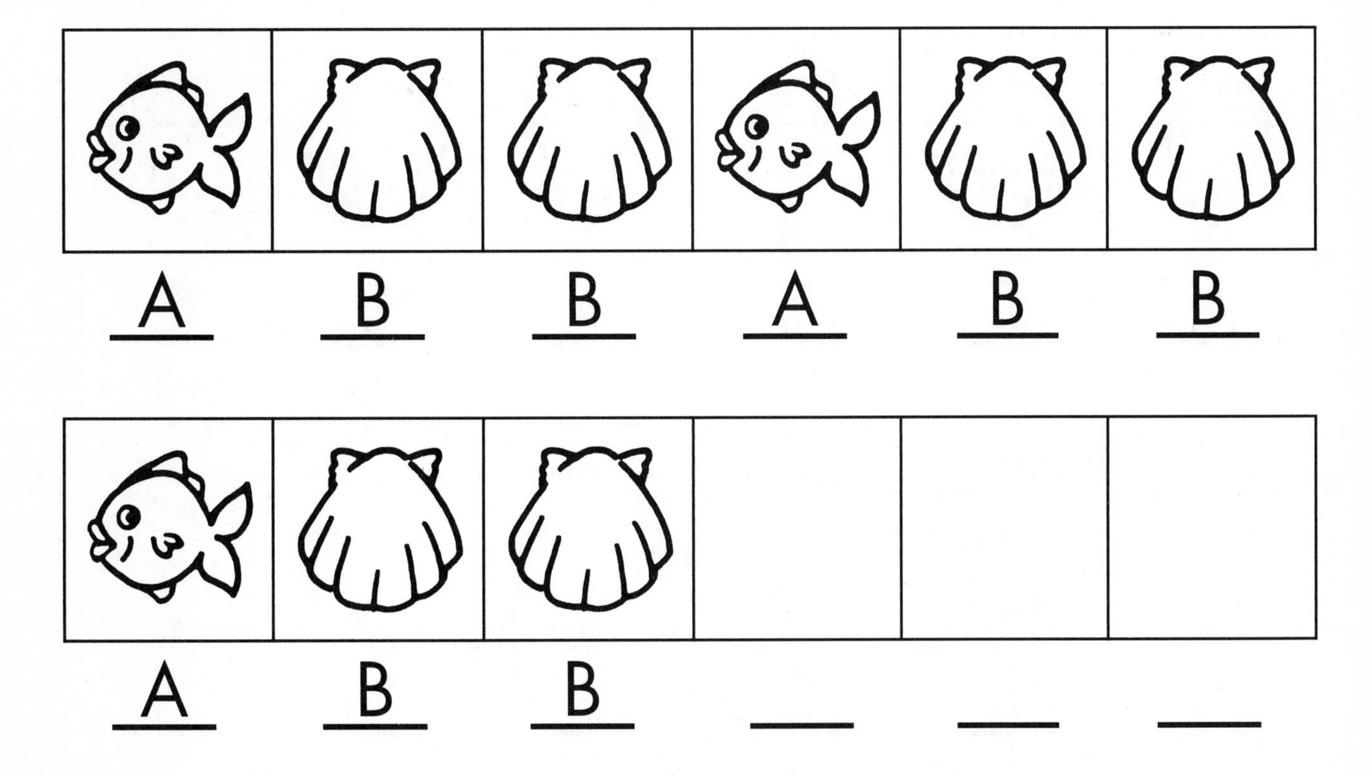

Extend the pattern!

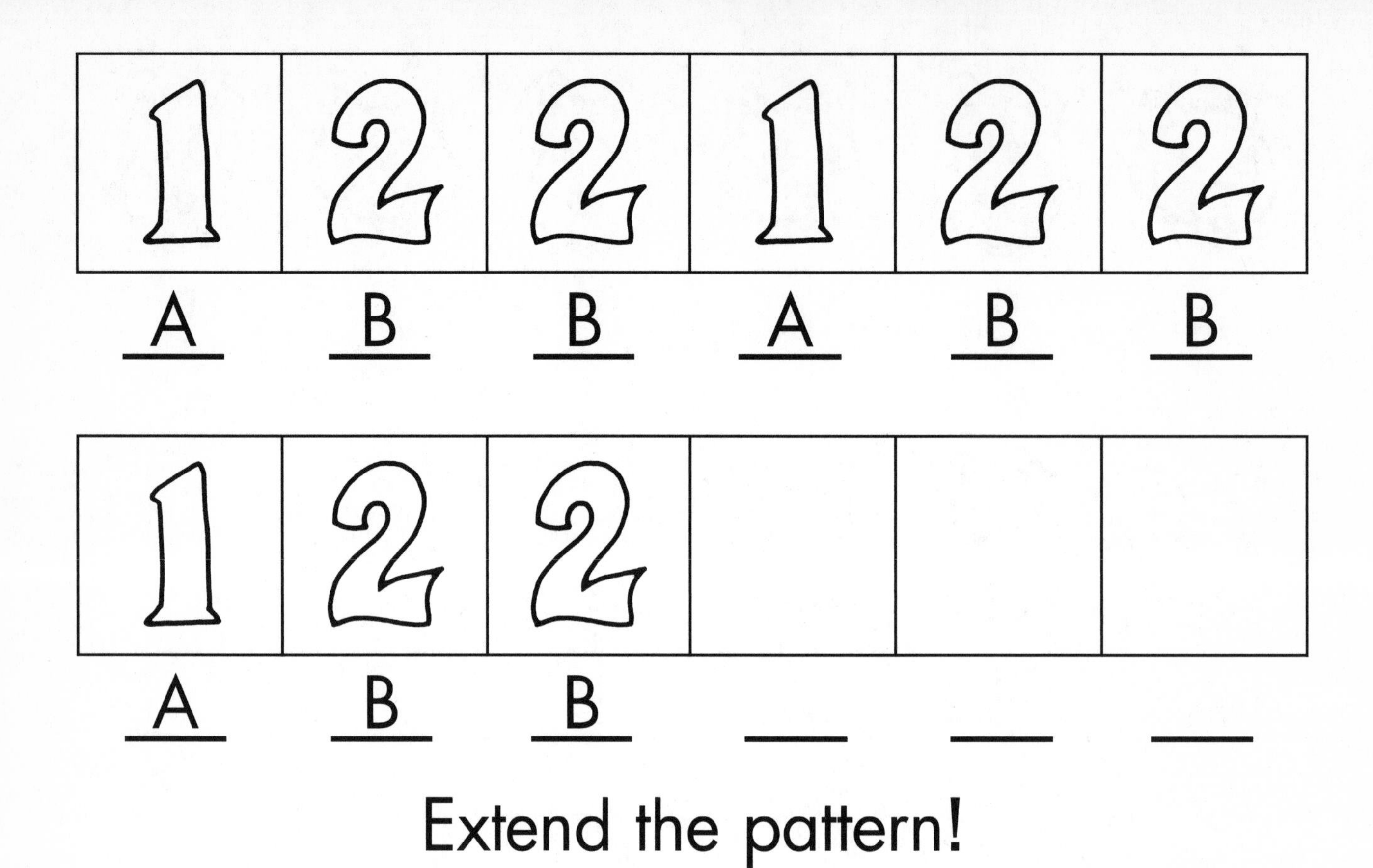

Extend the pattern!

Extend the pattern!

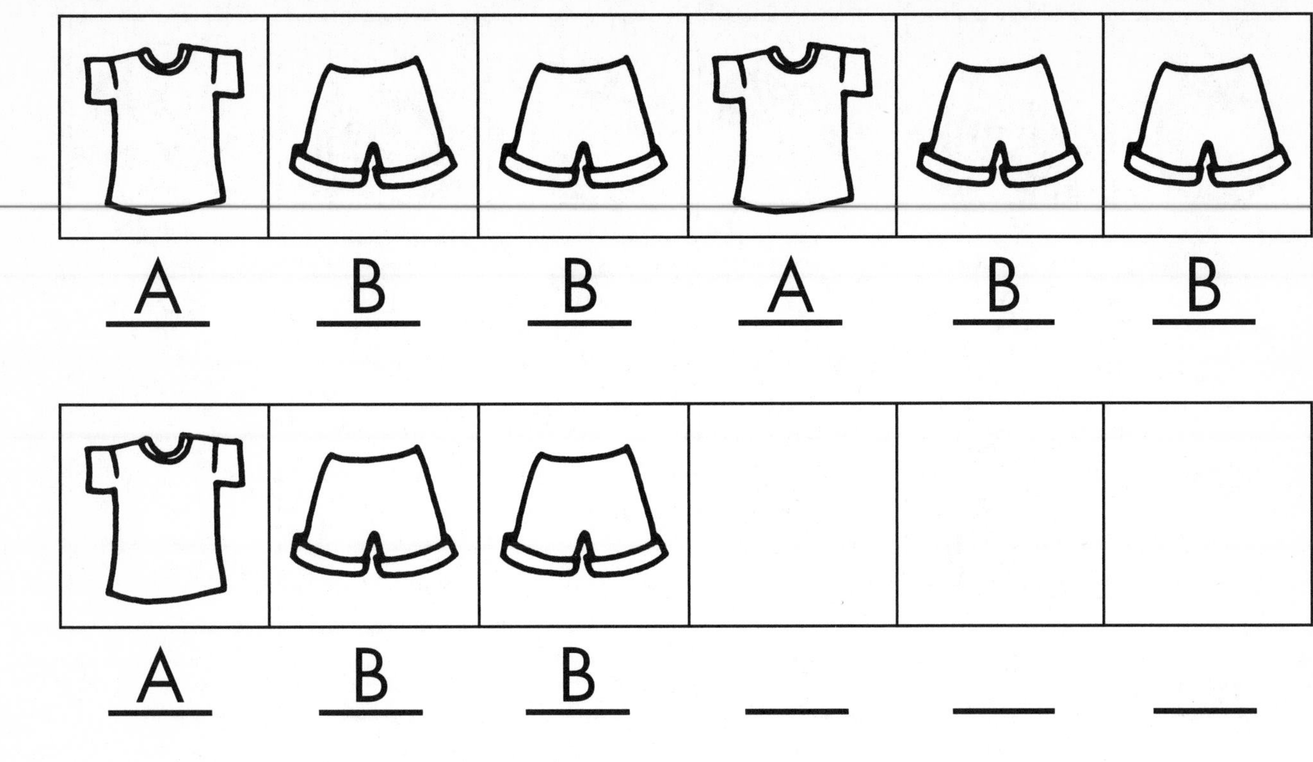

Extend the pattern!

5

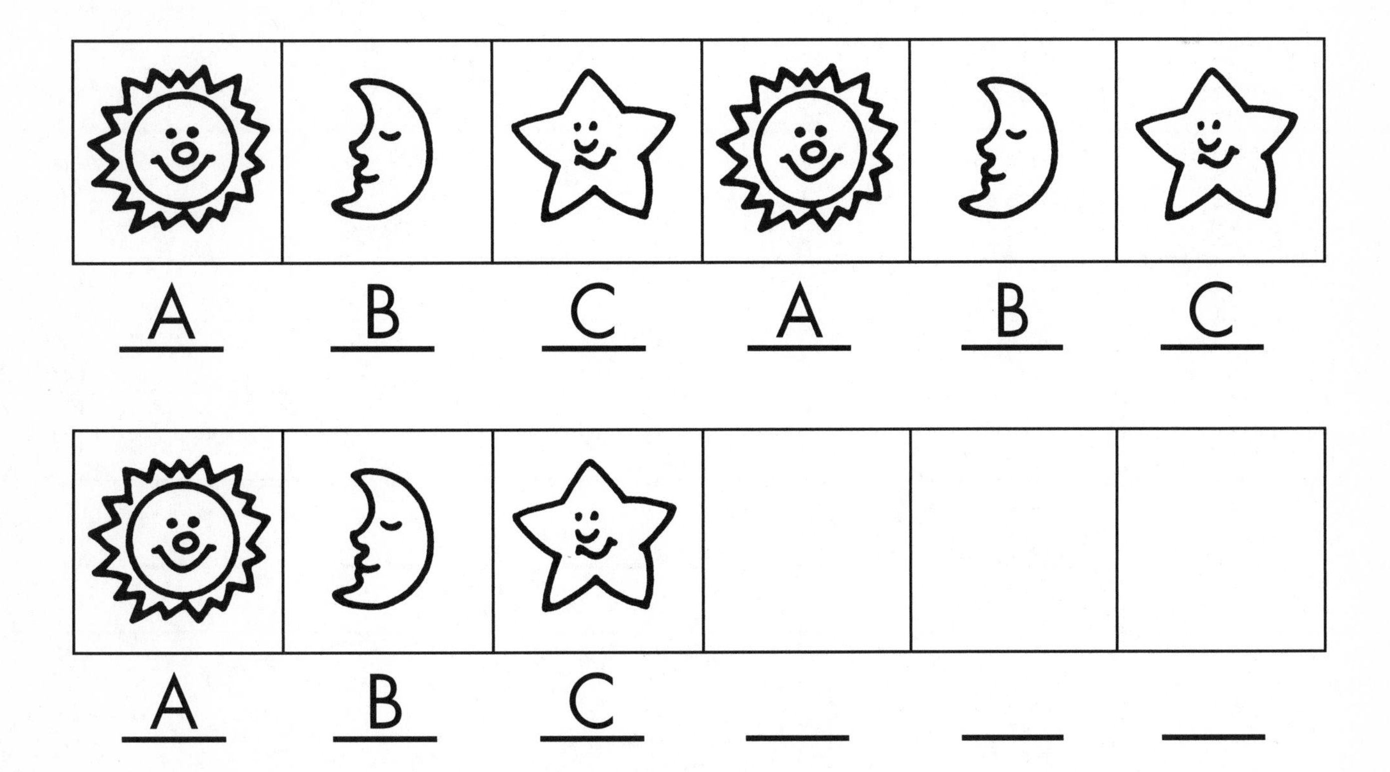

Extend the pattern!

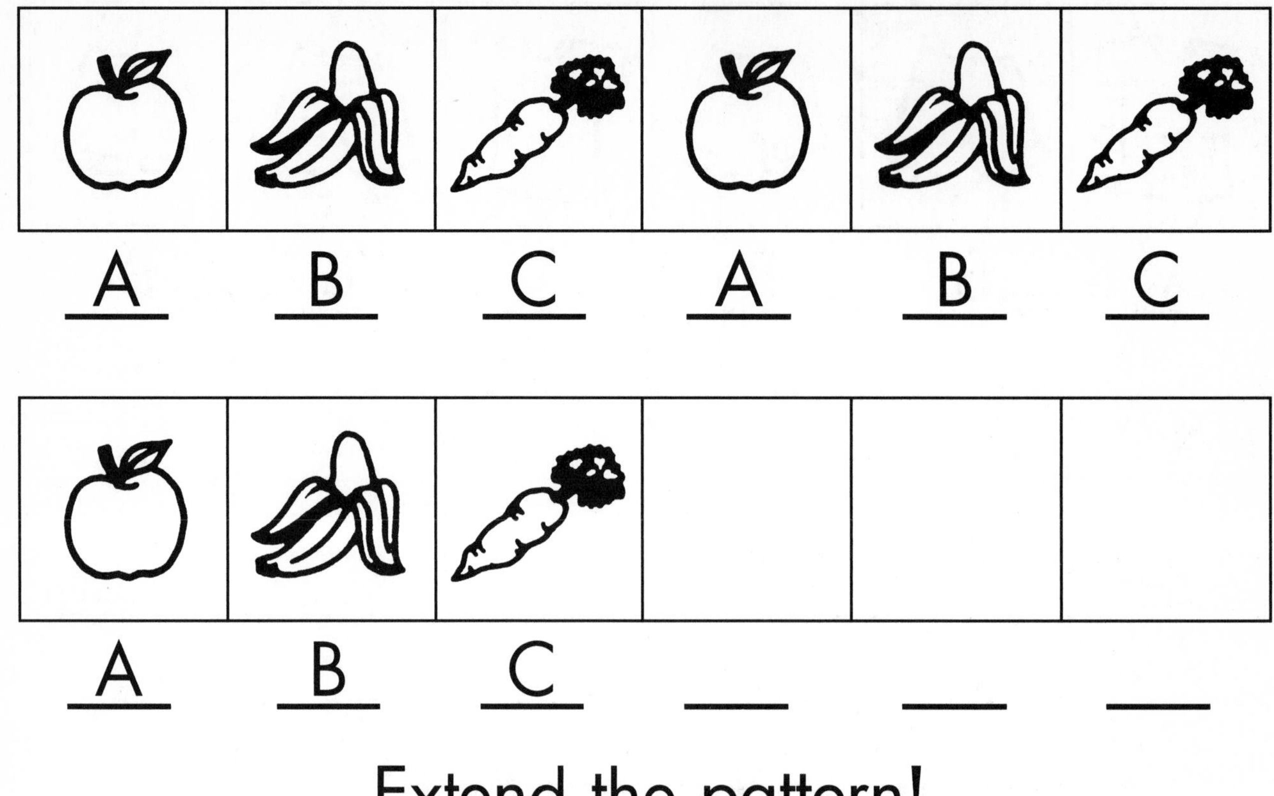

Extend the pattern!

A	B	B	___	___	___

I can make my own ABB pattern!

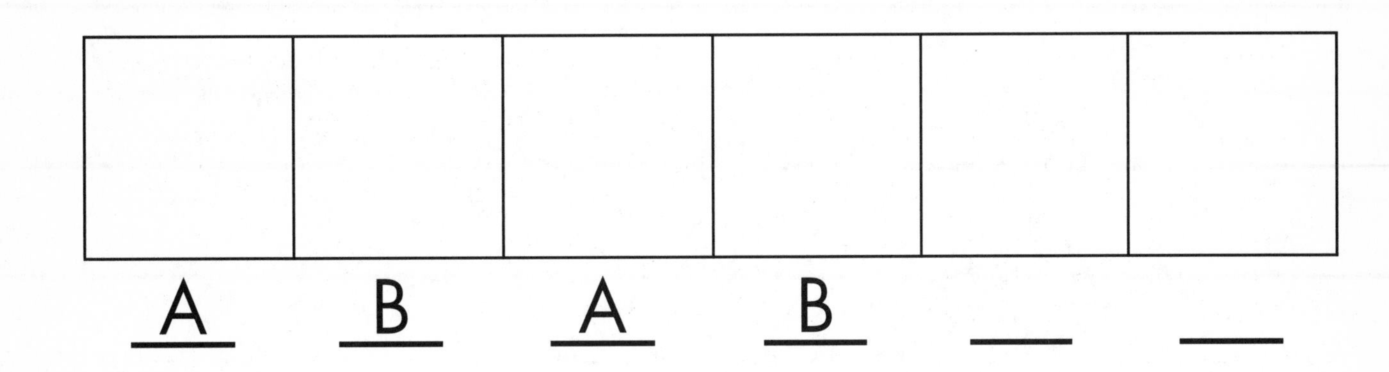

I can make my own AB pattern!

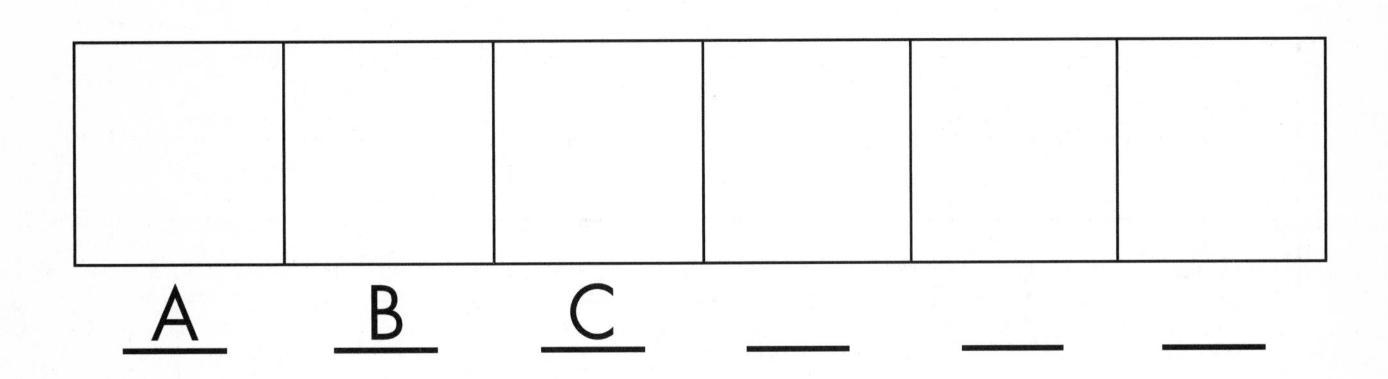

I can make my own ABC pattern!

Graph It!

by ____________________

How Many Animals Live on the Farm?

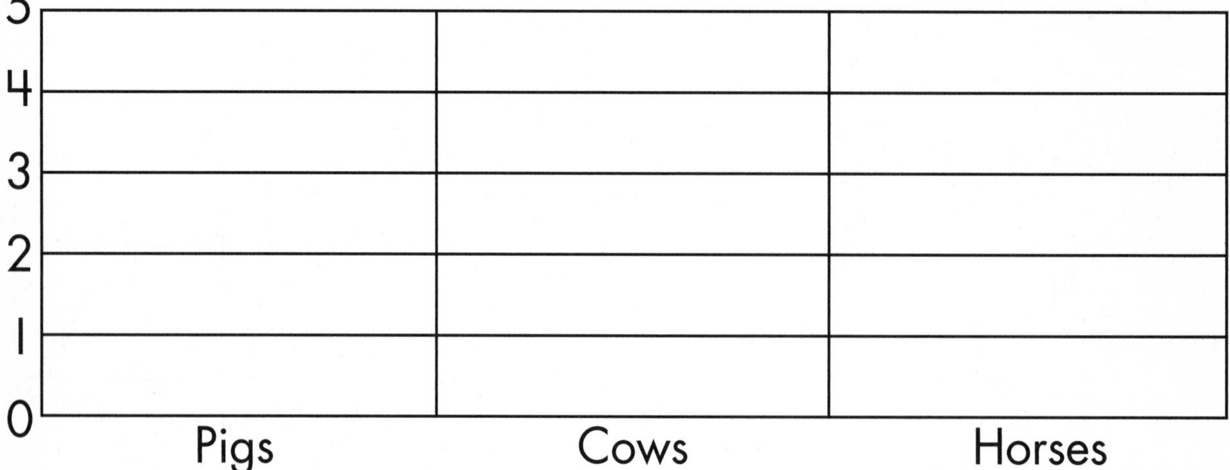

How many pigs live on the farm? ______

How many horses live on the farm? ______

How many cows live on the farm? ______

Read & Solve Math Mini-Books Scholastic Teaching Resources

Some animals live on a farm.

Some animals live at a zoo.

Read & Solve Math Mini-Books Scholastic Teaching Resources

How Many Animals Live at the Zoo?

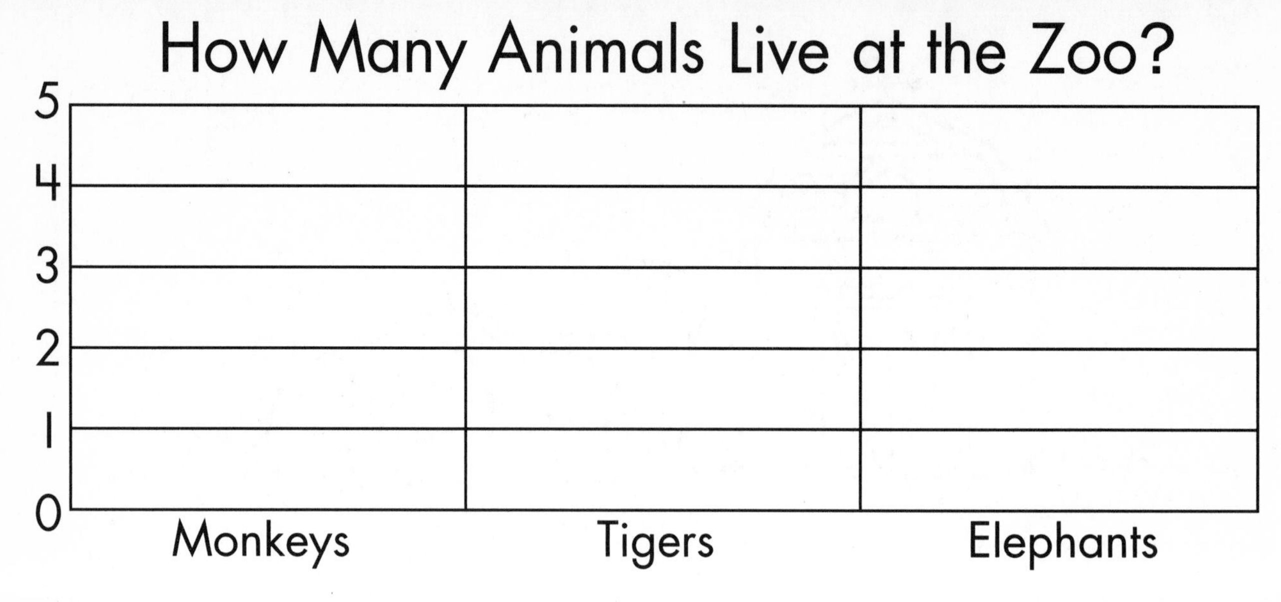

How many monkeys live at the zoo? ______

How many tigers live at the zoo? ______

How many elephants live at the zoo? ______

How Many Animals Live in the Sea?

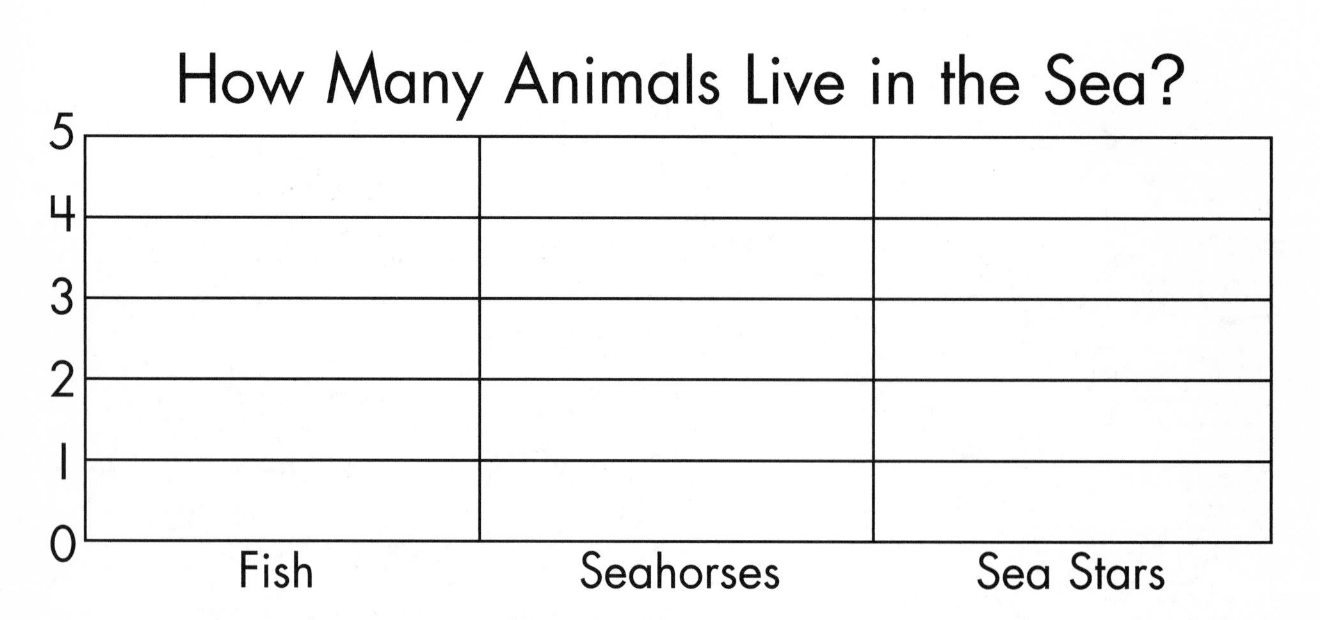

How many fish live in the sea? ______

How many seahorses live in the sea? ______

How many sea stars live in the sea? ______

Some animals live in the sea.

Some animals live by a pond.

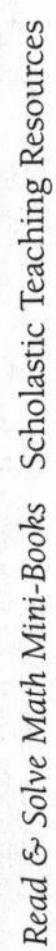

How Many Animals Live by the Pond?

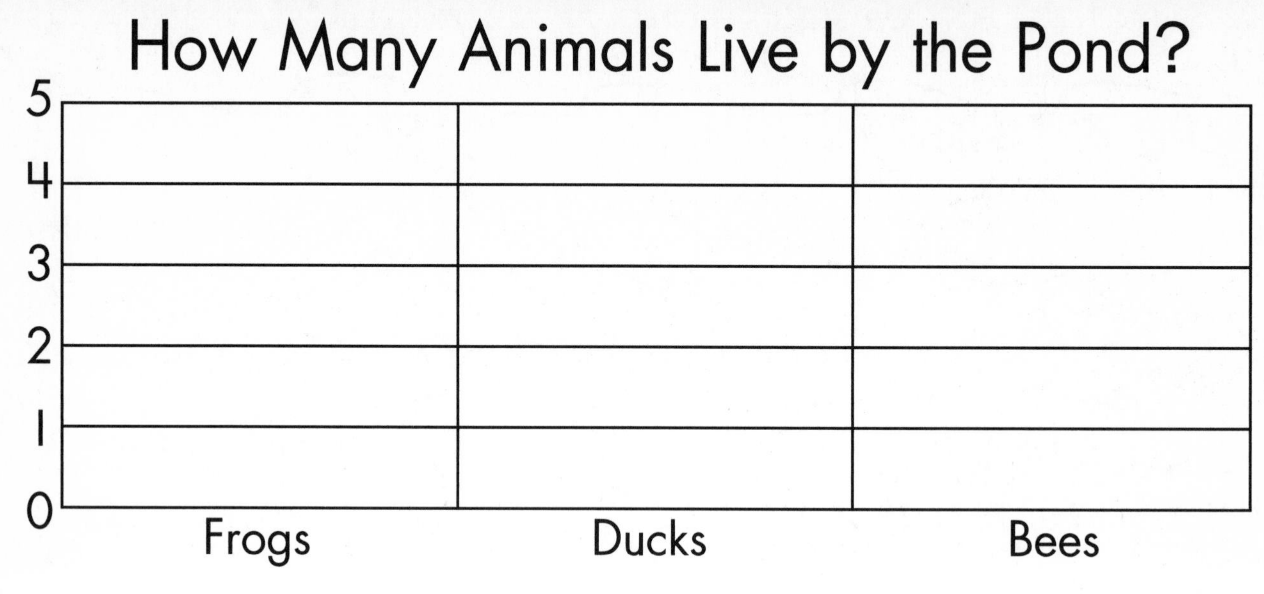

How many frogs live by the pond? ______

How many ducks live by the pond? ______

How many bees live by the pond? ______

How Many Animals Live in the Garden?

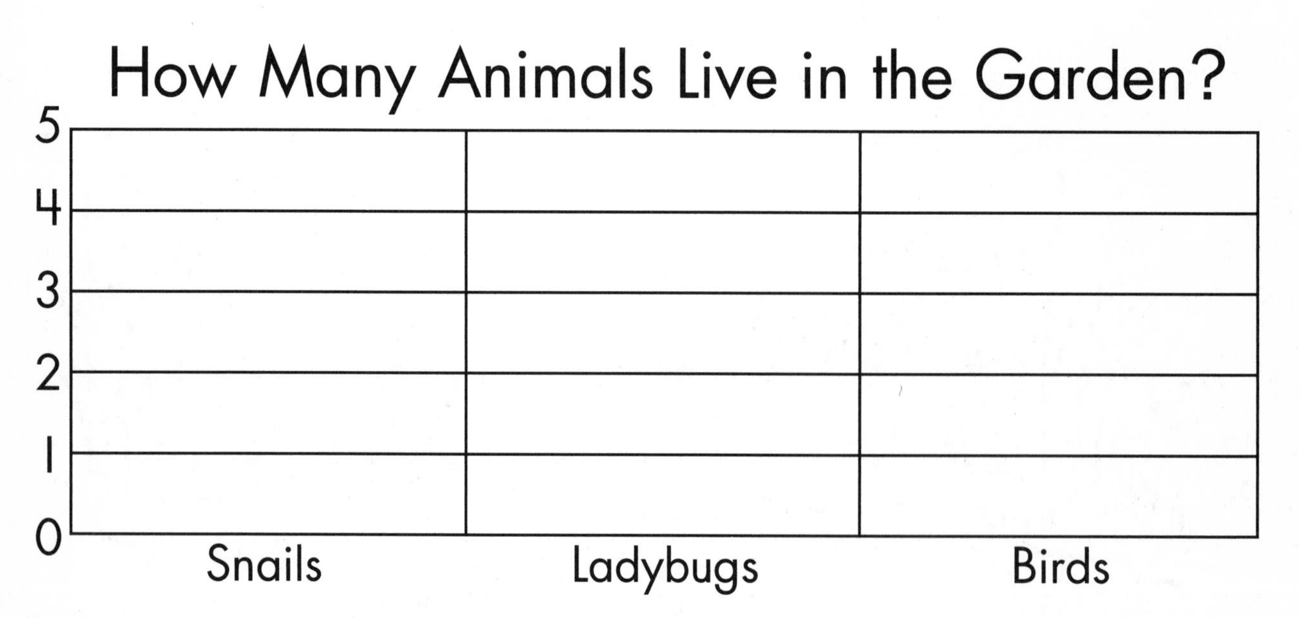

How many snails live in the garden? ______

How many ladybugs live in the garden? ______

How many birds live in the garden? ______

Some animals live in a garden.

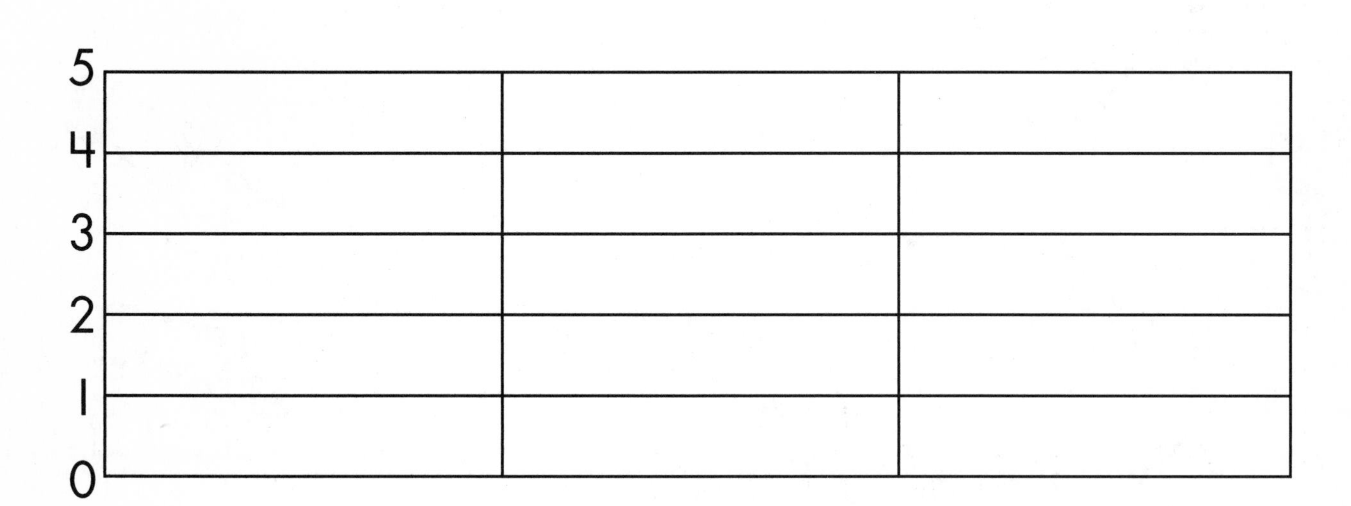

I can make my own graph!

Long and Short

by ______________________________

Add a long leash to the large dog.
Add a short leash to the small dog.

Add a long string to the large kite.
Add a short string to the small kite.

Add a long tail to the large mouse.
Add a short tail to the small mouse.

Add a long tongue to the large frog.
Add a short tongue to the small frog.

Add a long fishing line to catch the large fish.
Add a short fishing line to catch the small fish.

Add a long stem to the large flower.
Add a short stem to the small flower.

Add a long stream of exhaust to the large jet.
Add a short stream of exhaust to the small jet.

Around the Clock

by ______________________________

At nine o'clock, I go to school.

At eight o'clock in the morning, I wake up.

1

At ten o'clock, we read.

At eleven o'clock, we have math.

At one o'clock, we write.

At twelve o'clock, we eat lunch.

At two o'clock, we have recess.

At three o'clock, I go home.

At six o'clock, I eat dinner.

At four o'clock, I do my homework.

At eight o'clock at night, I go to bed.

12
11 1
10 2
9 3
8 4
7 5
6

We Eat Through the Week!

by ______________________________

Monday

Monday is the ________ day of the week.
On Monday we eat meatballs.

Sunday is the ________ day of the week.
On Sunday we eat spaghetti.

Tuesday

Tuesday is the ________ day of the week.
On Tuesday we eat tuna.

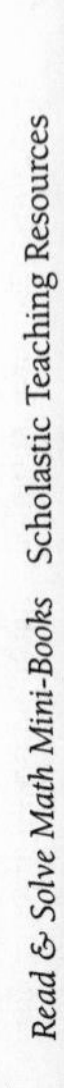

Wednesday

Wednesday is the ________ day of the week.
On Wednesday we eat waffles.

Friday

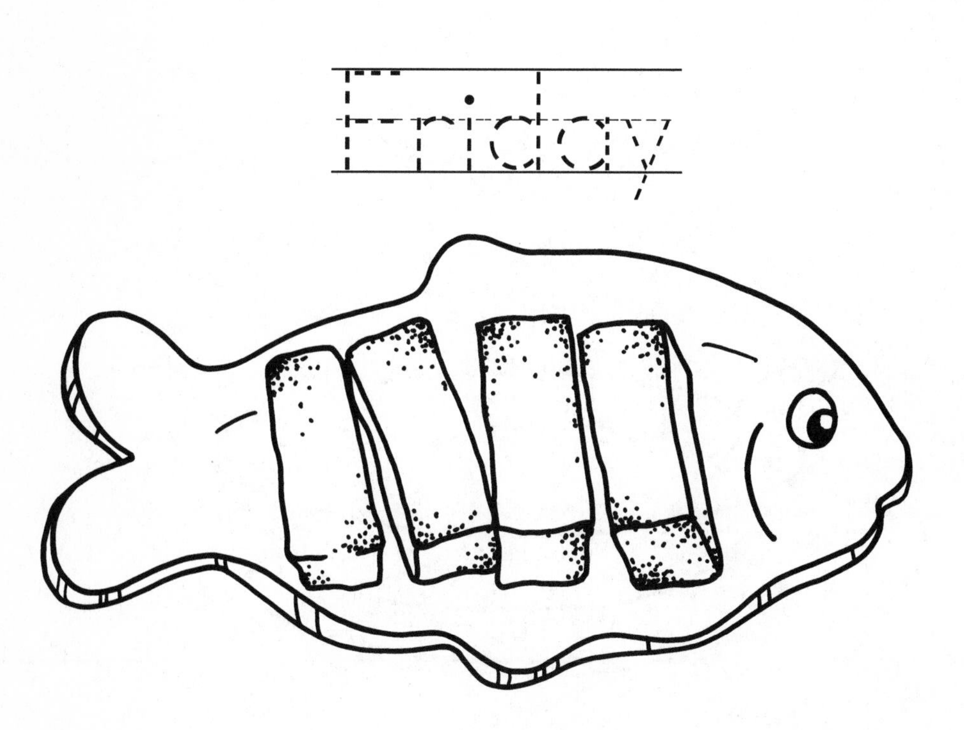

Friday is the ________ day of the week.
On Friday we eat fish sticks.

Thursday

Thursday is the ________ day of the week.
On Thursday we eat three-bean salad.

Saturday

Saturday is the ________ day of the week.
On Saturday we eat soup.